New Directions in the Philosophy of Science

Series Editor: **Steven French**, Professor of Philosophy of Science,
University of Leeds, UK

The philosophy of science is going through exciting times. New and productive relationships are being sought with the history of science. Illuminating and innovative comparisons are being developed between the philosophy of science and the philosophy of art. The role of mathematics in science is being opened up to renewed scrutiny in the light of original case studies. The philosophies of particular sciences are both drawing on and feeding into new work in metaphysics, and the relationships between science, metaphysics, and the philosophy of science in general are being re-examined and reconfigured.

The intention behind this new series from Palgrave Macmillan is to offer a new, dedicated publishing forum for the kind of exciting new work in the philosophy of science that embraces novel directions and fresh perspectives.

To this end, our aim is to publish books that address issues in the philosophy of science in the light of these new developments, including those that attempt to initiate a dialogue between various perspectives, offer constructive and insightful critiques, or bring new areas of science under philosophical scrutiny.

Titles include:

THE APPLICABILITY OF MATHEMATICS IN SCIENCE
Indispensability and Ontology
Sorin Bangu

THE PHILOSOPHY OF EPIDEMIOLOGY
Alex Broadbent

PHILOSOPHY OF STEM CELL BIOLOGY
Knowledge in Flesh and Blood
Melinda Fagan

INTERPRETING QUANTUM THEORY
A Therapeutic Approach
Simon Friederich

SCIENTIFIC ENQUIRY AND NATURAL KINDS
From Planets to Mallards
P. D. Magnus

COMBINING SCIENCE AND METAPHYSICS
Contemporary Physics, Conceptual Revision and Common Sense
Matteo Morganti

COUNTERFACTUALS AND SCIENTIFIC REALISM
Michael J. Shaffer

ARE SPECIES REAL?
An Essay on the Metaphysics of Species
Matthew Slater

MODELS AS MAKE-BELIEVE
Imagination, Fiction and Scientific Representation
Adam Toon

Forthcoming titles include:

SCIENTIFIC MODELS AND REPRESENTATION
Gabriele Contessa

New Directions of the Philosophy of Science
Series Standing Order ISBN 978–0–230–20210–8 (hardcover)

(outside North America only)

You can receive future titles in this series as they are published by placing a standing
order. Please contact your bookseller or, in case of difficulty, write to us at the address
below with your name and address, the title of the series and the ISBN quoted above.

Customer Services Department, Macmillan Distribution Ltd, Houndmills, Basingstoke,
Hampshire RG21 6XS, England

Interpreting Quantum Theory

A Therapeutic Approach

Simon Friederich

First published 2015 by
PALGRAVE MACMILLAN

Palgrave Macmillan in the UK is an imprint of Macmillan Publishers Limited,
registered in England, company number 785998, of Houndmills, Basingstoke,
Hampshire RG21 6XS.

Palgrave Macmillan in the US is a division of St Martin's Press LLC,
175 Fifth Avenue, New York, NY 10010.

Palgrave Macmillan is the global academic imprint of the above companies
and has companies and representatives throughout the world.

Palgrave® and Macmillan® are registered trademarks in the United States,
the United Kingdom, Europe and other countries.

ISBN 978-1-349-49619-8   ISBN 978-1-137-44715-9 (eBook)
DOI 10.1057/9781137447159

This book is printed on paper suitable for recycling and made from fully
managed and sustained forest sources. Logging, pulping and manufacturing
processes are expected to conform to the environmental regulations of the
country of origin.

A catalogue record for this book is available from the British Library.

A catalog record for this book is available from the Library of Congress.

Transferred to Digital Printing in 2015

To Sibylla, Alma, and Lydia

Contents

Part III Objections

Part IV Non-locality, Quantum Field Theory, and Reality

Series Editor's Foreword

The motivation behind our series is to create a dedicated publishing forum for the kind of exciting new work in the philosophy of science that embraces novel directions and fresh perspectives. To this end, our aim is to publish books that address issues in the philosophy of science in the light of these new developments, including those that attempt to bring new areas of science under philosophical scrutiny, initiate a dialogue between competing perspectives, or explore and develop distinctive new approaches.

Simon Friederich has written an exciting and provocative book that tackles the so-called measurement problem in quantum physics in a new and thought-provoking way. His overall philosophical stance is 'therapeutic', in a Wittgensteinian sense, whereby foundational problems are effectively dissolved by pointing out that they arise from certain conceptual confusions. In this particular case, Friederich argues that the confusion surrounds the notion of 'state' in quantum mechanics. Standardly, this term is taken to represent an objective feature of quantum systems, thus generating the problem of how to account for what we observe at the 'everyday' level given the 'entangled' states presented by quantum mechanics. If we drop this standard understanding and regard such states as reflecting our epistemic condition, so that the distinction between the quantum level and the everyday can be seen as a shift in our epistemic situation, then, Friederich insists, the problem simply dissolves.

Crucial to this approach is Searle's notion of a 'constitutive rule', in the sense of a rule that constitutes a particular activity (as in the case of the game of chess, for example). According to Friederich, to assign a state to a physical system in accordance with the rules of quantum mechanics is simply what it means to perform such a state assignment correctly, so that this assignment is logically dependent on those rules.

This 'Rule Perspective' offers an entirely novel stance on the foundations of quantum physics and Friederich sets it in the context of a carefully articulated review of both the relevant formalism and the most well-known interpretations of the theory, including the currently much-discussed 'Many Worlds View'. He also contrasts it with 'quantum Bayesianism', which is found to be wanting insofar as it cannot account for our knowledge of observables. A detailed account of how probabilities should be understood in quantum mechanics is then developed from the standpoint offered by the 'Rule Perspective'.

Friederich also carefully considers a range of possible objections to his approach. One is that it is just the infamous Copenhagen interpretation re-heated, but he notes the differences, as well as similarities, with the views of Bohr and Heisenberg. Another has to do with accommodating the extensive empirical success of quantum theory and Friederich shows how this can also be accounted for within his epistemic view. Finally, he tackles Bell's charge that such views amount to a form of anthropocentrism by articulating which uses of anthropocentric notions are legitimate in this context.

The book concludes by examining the implications of this approach for three further issues: non-locality, the foundations of quantum field theory, and scientific realism. In his final words, Friederich insists that it is not only *what* we learn from quantum theory about reality that is intriguing and fascinating but also *how* we do so.

Friederich's book represents a bold new approach to an old and much-discussed problem. It offers clearly articulated and powerful arguments for the dissolution of this problem and a re-orientation of our view of quantum physics in general. In this way it takes the field in a new and exciting direction. It is precisely the sort of work that the editorial board and I aim to support through the *New Directions* series and we are sure that it will have a major impact on both philosophers of science and philosophically reflective physicists.

Steven French
Professor of Philosophy of Science
University of Leeds

Preface

This book develops and explores what I call a 'therapeutic' approach to the foundational problems of quantum theory. It considers and develops the idea that quantum states do not represent any objective features of the quantum systems they are assigned to without denying that there are objective standards of correctness that govern the application of quantum theory to the world.

All the essential ingredients of my work on quantum theory in the past few years are combined (and further developed) in this work. Trying to formulate an account in the 'therapeutic' spirit just mentioned I have profited greatly from enormously helpful suggestions by many colleagues and friends. In particular, I would like to thank Andreas Bartels, who supervised my work on a thesis that eventually turned into this book. I am also grateful to Jeremy Butterfield, Michael Esfeld, Robert Harlander, Richard Healey, Koray Karaca, Felix Mühlhölzer, Thorben Petersen, Patrick Plötz, Gregor Schiemann, and Chris Timpson for the lively and inspiring exchanges I had with them on some of the topics treated here. Further useful comments on earlier versions of (what developed into) parts of this book were provided by anonymous referees of various journals and of Palgrave Macmillan. I would also like to thank the participants of the Göttinger Philosophisches Reflektorium as well as audiences in Heidelberg, Wuppertal, Bern, Göttingen, Bonn, Cambridge, Hanover, and Munich for numerous (partly very critical) valuable comments and suggestions. Most of all, I would like to thank my wife Andrea Harbach for her encouragement and our continuing marital dialogue.

While working on the 'therapeutic' approach to quantum theory presented here I had the immeasurable luck of welcoming my beloved daughters Sibylla, Alma, and Lydia (in order of appearance) to their lives. They have made my life so much richer while I have been working on this book, which is why I dedicate it to them.

Acknowledgements for permission

Some of the chapters of this book contain more refined and developed versions of earlier works. I acknowledge permission to reuse material that appeared for the first time in:

- How to spell out the epistemic conception of quantum states. *Studies in History and Philosophy of Modern Physics*, 42:149–157, 2011.
- In defence of non-ontic accounts of quantum states. *Studies in History and Philosophy of Modern Physics*, 44:77–92, 2013.
- Quantum theory as a method: The Rule Perspective. In T. Sauer and A. Wüthrich, editors, *New Vistas on Old Problems: Recent Approaches to the Foundations of Quantum Mechanics*, pages 121–138. Edition Open Access. Max Planck Research Library for the History and Development of Knowledge, 2013.
- Interpreting Heisenberg interpreting quantum states. *Philosophia Naturalis*, 50:85–114, 2013.
- Pristinism under pressure: Ruetsche on the interpretation of quantum theories. *Erkenntnis*, 78:1205–1212, 2013.

Part I

Introduction and Background

1
Introduction

1.1 Quantum foundations

Quantum theory is perhaps the theory with the greatest predictive and explanatory success in all the history of physics. To name just a few of its countless achievements: it explains the stability of the stable atoms and nuclei and predicts the decay rates of the unstable ones; it accounts for the different manifestations of matter such as gaseous, liquid and solid, metallic and insulating, magnetic, superfluid and superconducting; it forms the basis of our chemical knowledge; and it provides the conceptual framework of all contemporary models for the fundamental constituents of matter. In addition, some of its predictions, for example that of the 'electron spin g-factor' in quantum electrodynamics, are perhaps the most accurate ones ever made in the history of science. Some of its predictions concern matter at extremely high energy and interactions between bits of matter at extremely short distances; others are about matter at extremely low energy and close to the zero of absolute temperature. As far as its technological applications are concerned, its insights are at the heart of the overwhelming technological progress in information technology in the past few decades, it explains why nuclear fission and fusion work as they do, and it forms the basis of the rapidly growing field of nanotechnology.

However, all of quantum theory's stupendous predictive and explanatory achievements notwithstanding, the debates about its foundations are more hotly contested than ever. In fact, the majority view among philosophers of physics, accurately summarised by David Wallace, is that quantum theory 'is the great scandal of physics; despite [its] amazing successes, we have no satisfactory physical theory at all – only an ill-defined heuristic which makes unacceptable reference to primitives

3

such as "measurement", "observer" and even "consciousness"' (Wallace [2008], p. 16). Most philosophers of physics would agree that quantum theory either needs to be modified and replaced by a different theoretical framework, or that it requires an interpretation which leads to a picture of the world that is radically at odds with our everyday views. The most prominent examples of theories designed to replace quantum theory are pilot wave theory ('Bohmian mechanics') and GRW theory. The most famous speculative interpretations, which lead to a radical breaking with our everyday views, are the variants of the Everett interpretation, including the so-called many-worlds and many-minds interpretations. Their core idea is that our world is (or our minds are) constantly subject to ubiquitous branching processes, and that what we take to be the actual world (or our actual mind) is just one among countless many world (or mind) branches.

A few decades ago, the debate about the foundations of quantum theory was still shaped by widespread commitment to the so-called *Copenhagen interpretation*. Whatever exactly this interpretation actually says, one of its few clear-cut tenets (perhaps the only one) is that quantum theory is *complete* as an internally sound physical theory and requires neither *modification* nor *re-interpretation* in speculative terms. Almost all philosophers of physics would now reject this view.[1] Their main reason for regarding the Copenhagen interpretation as untenable (whatever exactly it says) can be summarised roughly as follows:

The Copenhagen interpretation must either embrace the 'collapse of the wave function' as one of quantum theory's crucial constituents (for introductory comments on the collapse of the wave function see Section 2.2) or reject it.[2] If it rejects collapse, the Copenhagen interpretation cannot account for the manifest – and empirically trivial – fact that measurement processes always result in determinate outcomes. It is then empirically inadequate in the most blatant way possible. If, in contrast, the Copenhagen interpretation accepts collapse, it must specify under which conditions collapses occur, but without resorting to such vague, ambiguous and anthropocentric notions as 'measurement' and 'observer'. These are inadequate in the context of supposedly fundamental physical theories and may therefore not be used to specify the criteria for the occurrence of collapse. Since neither Bohr nor Heisenberg nor Pauli nor Dirac nor any other supposed adherent of the Copenhagen interpretation seems to have proposed a coherent alternative to this fatal dilemma, it follows that the Copenhagen interpretation – whatever it actually says – is deeply unsatisfying; or so contemporary philosophers of physics essentially seem to agree.

The difficulty of accounting for determinate measurement outcomes in quantum theory is an aspect of the famous *measurement problem*. Almost all suggested modifications and interpretations of quantum theory can be seen as attempts to *solve* the measurement problem. They are naturally categorised (as I will show in Chapter 3) according to how they react to this problem.

The measurement problem is not the only substantial foundational challenge in quantum theory. Another arises from what is widely known as 'quantum non-locality': the fact that quantum theory predicts correlations between measurement outcomes which obtain independently of the spatio-temporal distance between the outcomes and which it does not explain in terms of *common causes*. Bell's theorem (for a sketch see Appendix A) shows that these correlations are incompatible with theories that respect a criterion that Bell takes to incorporate the idea of '*local causality*': that causal influences cannot travel faster than light and probabilities depend only on what occurs in regions from which influences at velocities no larger than the velocity of light can arrive. Superluminal influences are widely regarded as problematic in the relativistic context, mainly due to the fact that they are *backwards in time* in some inertial frames.

Thus, in view of Bell's result, quantum theory and relativity theory seem to clash at least in spirit (if not in letter) in a very profound and seemingly unavoidable way. Since the correlations predicted by quantum theory are experimentally well-confirmed, the pressure, according to many authors, is on relativity theory here. Some authors go as far as considering the revival of a space-time framework which contains a universally privileged inertial rest frame in terms of which an absolute simultaneity relation among space-time events would be defined.[3] If this conclusion were the right one to draw, one of the main lessons of relativity theory – that all inertial frames are on an equal conceptual footing – would have to be unlearned, and the symmetries that are regarded as principled in relativity theory would be downgraded to merely apparent and emerging rather than fundamental.

In conclusion, it appears that reflections on the foundations of quantum theory have the potential to undermine our confidence not only in quantum theory itself, but also in relativity theory. So, these reflections shake our faith in the two main elements of our contemporary understanding of physics without providing any clear ideas as to what should replace them. Is there any way to avoid this devastating result?

1.2 The idea of a therapeutic approach

The aim of the present work is to probe a specific way of answering this question with a 'yes, it can be avoided'. The hypothesis that I wish to explore is that the foundational problems require neither revising quantum theory or relativity theory nor any extravagant metaphysics. The idea which motivates this investigation is that both the measurement problem and the problem of quantum non-locality are mere artefacts of conceptual confusions that disappear once the actual role of the elements of the quantum theoretical formalism in its applications is properly taken into account. I refer to this type of approach as 'therapeutic', as it tries to 'cure' us from what it takes to be unfounded worries that arise from conceptual confusion.

The idea of philosophy as a form of intellectual therapy has a long tradition (one may trace it back to Epicurus in antiquity). In the more recent history of philosophy it is advocated in particular by the later Wittgenstein. The therapeutic approach to philosophical problems is a core element of the radically innovative conception of philosophy he develops and puts to work in his *Philosophical Investigations* (Wittgenstein [1958]). Wittgenstein's suggested strategy for dissolving philosophical problems is to reflect on the actual context and mode of use of the concepts involved, i.e. their roles in our 'language games', and to investigate whether, in order to formulate the problems, they are perhaps used in ways in which they cease to make good sense. Typically, the Wittgensteinian diagnosis is that the source of a perceived philosophical problem is that one had naively – and mistakenly, as it turns out – assumed that the concepts involved in its formulation are used descriptively (or, more specifically, as more or less directly representing features of reality), which in fact they are not.[4] Examples of Wittgensteinian therapeutic analyses include his account of mental concepts such as 'pain' (see, e.g., (Wittgenstein [1958] § 244)) and his account of mathematical language as normative rather than descriptive.[5] In both cases, his ambition is to make the philosophical problems associated with these concepts disappear by constantly reminding us of their actual, 'everyday' (Wittgenstein [1958] § 116), use.

The present work applies the idea of philosophy as therapy to the foundational problems of quantum theory. With respect to the measurement problem, the most straightforward strategy for dissolving it is to suggest that the core quantum theoretical vocabulary – namely, quantum states – is actually used non-representationally, contrary to what is usually assumed. Whether or not this view was historically held by those associated with the Copenhagen interpretation (a matter that will be discussed in Chapter 7), it is arguably what their view as regards quantum states *should* have been. Certainly the most straightforward advantage of regarding quantum states as not representing any objective features of the systems they are assigned to, but instead as somehow reflecting the epistemic conditions of those who assign them, is that it permits an unproblematic interpretation of measurement collapse as reflecting a sudden *change* in the epistemic situation of the agent assigning the quantum state.

In this work I develop what seems to me the most attractive way of spelling out the idea that quantum states reflect the epistemic conditions of the agents who assign them. The account which I propose is partly based on ideas drawn from Richard Healey's recent pragmatist interpretation (Healey [2012a]), and it partly builds on my own ideas as to how Searle's notion of a *constitutive rule*, as suggested by (Searle [1969]), may be usefully put to work in the quantum foundational context.

My conclusion as regards the prospects of avoiding the measurement problem along therapeutic lines is tentatively optimistic: the most prevalent criticisms against the viability of interpretation of quantum states as partly epistemic can be answered, and the charge of instrumentalism does not apply. The old Copenhagenian stance that quantum theory does not necessarily need either modification or speculative reinterpretation can be vindicated. Nevertheless, the project of assigning determinate, sharp values to a large class of observables remains a live one: based on the considerations to be developed in the meantime and nourishing the hopes of those who are looking for a 'robustly realist' interpretation, Chapter 12 argues that the prospects for finding such assignments in a natural way may be much better than widely thought.

With respect to the problem of compatibility between quantum theory and special relativity the upshot of my considerations is especially encouraging: the upshot of the discussion of these matters to be given here is that once we adopt a perspective on 'causality' and 'probability' which properly takes into account the actual roles of these concepts, the apparent violation of local causality in quantum theory disappears and the putative tension between quantum theory and relativity theory is revealed to be spurious. In the next section, I give an outline of my main arguments for the claims just sketched by providing an overview of the structure of this book.

1.3   Outline of this work

The twelve chapters in this work are grouped together in four parts, each with three chapters. In addition to the present introduction to what follows, the three chapters of Part I contain a review of the quantum theoretical formalism and give an overview of the most-discussed interpretations of the theory. Part II motivates and develops an account of quantum theory – the 'Rule Perspective' – that is meant to dissolve these problems in the therapeutic vein sketched before. Part III discusses and answers objections that one may bring forward to challenge this account. Finally, Part IV discusses further relevant miscellaneous topics, namely, the relation between quantum theory and relativity theory, the bearing of quantum field theory on the discussion, and the status relation between quantum theory and 'reality' according to the account developed before. The remainder of this section gives a more detailed overview of the individual chapters.

In order to provide common ground for the discussions that will follow, Chapter 2 presents a rough sketch of the most elementary features of the quantum theoretical formalism: that it represents physical observables by self-adjoint operators on a Hilbert space $\mathcal{H}$, the quantum states (assignments of probabilities to the possible values of observables) by density matrices ρ on $\mathcal{H}$, and the probabilities of the possible values themselves by expectation values of projection operators on $\mathcal{H}$. The projection postulate and the quantum theoretical treatment of many-component quantum systems, including the famous concept of quantum entanglement, are also introduced.

Chapter 3 gives a comprehensive overview of the measurement problem. Following an exposition due to (Maudlin [1995]), it is presented as a conflict between the assumptions that (A) quantum states give complete descriptions of quantum systems, (B) the Schrödinger equation (which governs the time-evolution of quantum states) is universally valid, and (C) that measurements have determinate outcomes. *Solutions* to the measurement problem are approaches which reject at least one of these assumptions as false. Existing interpretations of quantum theory – inasmuch as they are solutions to the measurement problem – are reviewed and classified according to which of the assumptions (A), (B), and (C) they reject. This completes the first, predominantly introductory, Part I of the book.

Chapter 4, the first of Part II, starts to develop the account of quantum theory to be proposed here by introducing the concept of a *dissolution* of the measurement problem, contrasting it with that of a *solution*. Just like solutions, dissolutions also reject at least one of assumptions (A), (B), and (C), but for being senseless rather than for being wrong or, to put it more diplomatically, for being based on mistaken conceptual presuppositions rather than for simply being mistaken. I explain in which way the idea of dissolving the measurement problem implements the therapeutic approach to philosophical problems mentioned above and review the inspiration and origin of this approach in the philosophical method of the later Wittgenstein.

If the role of quantum states is not that of representing features of reality, then what is it? A natural conjecture is that quantum states somehow (in a way to be made more precise) reflect aspects of the epistemic situations of the agents assigning them. Indeed, on such an *epistemic* (as opposed to *ontic*) reading of quantum states, the measurement problem cannot even be formulated (which does not mean that it may not arise in a different form; see Chapter 9). But the question remains as to how an epistemic account of quantum states should be formulated in detail.

Chapter 5 approaches this task by reviewing quantum Bayesianism, which is presently the most-discussed epistemic account of quantum states. This radically subjectivist position conceives of quantum probabilities as subjective degrees of belief for which there is neither right nor wrong. Though plausible and inspiring in several respects, quantum Bayesianism falters due to its inability to account for the manifest fact that we are able to have knowledge of the values of observables. Having formulated this criticism, I propose and develop a novel approach to the epistemic conception of states that is based on Searle's notion of a *constitutive rule*. Due to the central role which this account ascribes to the rules that govern the assignment of quantum states, I refer to it as the 'Rule Perspective'.

Quantum theory is an essentially probabilistic theory. This raises the following two questions: first, what are the bearers of quantum probabilities? Second, are these probabilities subjective or objective? Chapter 6 provides answers to these questions from the point of view of the Rule Perspective.

The question concerning the bearers of quantum probabilities is answered by drawing on aspects of Richard Healey's recently proposed pragmatist interpretation of quantum theory: the bearers of probabilities are the so-called non-quantum magnitude claims ('NQMCs'), i.e. statements of the form 'the value of the observable A lies in the range Δ', which, however, as convincingly argued by Healey, are not all 'licensed' (in a sense to be discussed) for each system at all times.

The question concerning the nature of quantum probabilities is answered by saying that they are *objective* inasmuch as there is right and wrong in their ascription and *subjective* inasmuch as what the correct probabilities are depends on the evidence to which the agents who ascribe them have access. David Lewis' Principal Principle – which, roughly speaking, states that only that can be an objective probability what prescribes the rational credence for some (actual or hypothetical) agent – is recommended as a useful guide to the interpretation of quantum probabilities in the context of the Rule Perspective, smoothly connecting their objective and subjective aspects. Finally, an objection to interpretation of quantum probabilities as subjective (Timpson's so-called Quantum Bayesian Moore's Paradox) is considered and found not to arise as a problem for the Rule Perspective.

By suggesting that quantum theory is fine as it stands and does not need any interpretation in terms of hidden variables, spontaneous collapses, many worlds, and the like, the Rule Perspective is in at least one important respect similar to the famous (and notoriously elusive) Copenhagen interpretation. One may therefore be tempted to criticise the Rule Perspective for being a mere restatement of the views that were held by these eminent physicists but are nowadays widely regarded as obsolete. Part III of the book opens with Chapter 7 by trying to answer this question.

With respect to Bohr, giving a clear-cut answer turns out to be difficult, since Bohr's writings are elusive and notoriously hard to understand. One important difference with respect to Bohr concerns his insistence on the enduring importance of the language of classical physics, a claim which is not endorsed by the Rule Perspective. With respect to Heisenberg, there are more clear-cut similarities and differences. The most important similarity is that Heisenberg also insists on the interpretation of measurement collapse as an update in the agent's epistemic situation. The most important difference concerns his interpretation of quantum probabilities as 'objective tendencies', which is naturally read as excluding any view according to which quantum probabilities may be different for different agents.

As noted in the beginning of this chapter, quantum theory is one of the most impressive theories in the history of science as far as explanatory success is concerned, and any interpretation needs to account for how this explanatory success is possible. Addressing a challenge by Timpson against quantum Bayesianism, Chapter 8 considers the objection that the Rule Perspective, by relying on an epistemic account of quantum states, is unable to account for that explanatory success. The

challenge is rejected by outlining how both the evidential and causal
·aspects of quantum theoretical explanation can be accounted for by the
Rule Perspective.

While the Rule Perspective dissolves the measurement problem by
undermining the conceptual presuppositions of the above assumptions
(A), (B), and (C), it contains no criterion of under which conditions
which observables have determinate values. Furthermore, it uses anthropocentric notions (such as 'agent' and 'epistemic situation') to account
for in which cases quantum states are subjected to collapse. This re-
invites Bell's influential criticism of the Copenhagen interpretation and
related views for their unacceptable appeal to anthropocentric notions
in the interpretation of a supposedly fundamental theory.

Chapter 9 considers Bell's charge of anthropocentrism in detail and
answers it by carefully reflecting on which uses of anthropocentric
notions are legitimate in interpretations of fundamental physical theories and which are not. It concludes with an assessment of why
determinate measurement outcomes – far from being miraculous from
the point of view of the Rule Perspective – are precisely what competent
users of quantum theory should assume and expect.

Part IV covers three further topics of interest and relevance that are
subsumed under its title 'Non-locality, Quantum Field Theory, and Reality'. Chapter 10 starts by returning to the other main foundational
problem of quantum theory besides the measurement problem, namely,
the alleged deep-seated tension between quantum theory and relativity
which many believe to arise from 'quantum non-locality'. The idea that
there is such a tension rests partly on the view that quantum correlations require superluminal causal influences, which, however, is shown
to be contestable by investigating the counterfactuals to which those
correlations appear to give rise. The most widely shared worry as regards
the compatibility between quantum theory and relativity arises from
the fact that quantum correlations apparently violate Bell's probabilistic
criterion of 'local causality', which is meant to spell out the idea that
the probabilities of events depend only on what occurs in their backward light cones. According to Bell, a theory is locally causal if, for the
probability $Pr(A)$ of an event A, the identity

$$Pr(A|E) = Pr(A|E,B) \tag{1.1}$$

holds, where B is space-like separated from A, and E is a complete specification of what happens in A's backward light cone (the region from
where influences at velocities no larger than the velocity of light can
reach A).

The present work disputes the widespread belief that quantum theory is in conflict with the idea that Bell's criterion tries to make precise. Returning to the Lewisian insight, as encoded in the Principal Principle, that objective probability can only be what imposes constraints on rational credence, an argument is presented to the end that whether or not the condition $Pr(A|E) = Pr(A|E,B)$ holds is irrelevant to whether local causality, properly speaking, holds. I argue further that we have very good reasons for believing that quantum theory conforms to local causality – on a reading of that notion that takes the lessons of the Principal Principle into account. Furthermore, I offer a diagnosis of how the misleading impression that quantum and relativity theory are in tension arises.

Chapter 11 addresses some of the additional philosophical challenges which quantum *field* theory harbours beyond ordinary non-relativistic quantum theory for systems with finitely many degrees of freedom. For example, in the mathematically rigorous algebraic approach to quantum field theory, there is much debate as to *which* quantum states should be counted as physically relevant. In the literature on this topic, the debate has been framed as that between 'Hilbert Space Conservatism' and 'Algebraic Imperialism'. Recently, an interesting solution to the problem of deciding between those two (equally unattractive) positions has been proposed in Laura Ruetsche's book *Interpreting Quantum Theories* (Ruetsche [2011]). I consider Ruetsche's proposal and argue that it finds its natural home in the context of an epistemic account of quantum states.

Chapter 12, the final chapter of the book, concludes by asking to what degree the Rule Perspective commits one to an instrumentalist view of quantum theory and to what degree realism remains available. The chief non-realist element of the position, I contend, is that it conceives of quantum theory as unable, at least by itself, to provide a complete account of the history of the universe from an agent-independent perspective. Nevertheless, the assumption that agent-independent true statements can be made about physical reality in the form of NQMCs remains crucial and can be seen as the realist 'backbone' of the position. Moreover, as it turns out, there are reasons to be confident that even the ambitiously 'realist' project of assigning sharp values to all observables of a quantum system may be viable in a manner that is both natural and compatible with quantum theory as viewed from the standpoint of the Rule Perspective.

2
Sketch of the Formalism

This chapter reviews the quantum theoretical formalism and sketches its foundational problems. The main function of this chapter is to establish common ground for the interpretive discussions which follow, not to provide a comprehensive introduction to quantum theory.[6] Its first section recapitulates the essential aspects of the Hilbert space formalism of quantum mechanics, the second section covers the textbook account of measurement and measurement collapse, and the third and final section outlines the quantum theoretical treatment of many-component systems and explains the concept of quantum entanglement. Further introductory material is provided in the Appendices A, B, and C, all of which cover important *no-go* theorems for the interpretation of the theory.

2.1 The Hilbert space formalism

The physical objects to which quantum theory is applied are referred to as *quantum systems*. The physical properties of these systems – for example position, momentum, angular momentum, kinetic and potential energy – are conventionally referred to as *observables*. This terminology, also adopted here, is not meant to carry any commitment to actual observability, whether in practice or under suitably idealised conditions. Quantum theory ascribes expectation values to observables or, connectedly, probabilities to their possible values. The ascriptions of expectation values and probabilities are commonly – and perhaps somewhat misleadingly – referred to as 'quantum states'. In the formalism of quantum theory, the observables and the quantum states are expressed as follows:

The observables, to begin with, correspond to the self-adjoint elements $A = A^*$ of the algebra $\mathcal{B}(\mathcal{H})$ of bounded linear operators on a

Hilbert space $\mathcal{H}$. Typically, one arrives at a quantum theory by translating the *canonical structure* that some classical theory to be 'quantised' has in the Hamiltonian ('canonical') formalism into a *commutation* or *anti-commutation* relation among elements of the Hilbert space linear operators. The canonical structure of the classical theory is encoded in terms of the *Poisson bracket* relations between observables, where the Poisson bracket $\{\}$ is defined by $\{f,g\} = \sum_{i=1}^{N} \left[\frac{\partial f}{\partial q_i} \frac{\partial g}{\partial p_i} - \frac{\partial f}{\partial p_i} \frac{\partial g}{\partial q_i} \right]$. As a relation between position and momentum observables, the canonical structure relation in the Hamiltonian formalism reads

$$\{x_i, p_j\} = \delta_{ij}, \tag{2.1}$$

where the variables $i,j = 1,\ldots,N$ label the degrees of freedom of the system (assumed to be finite). Canonical quantisation of a classical theory promotes this relation to the so-called *canonical commutation relation* that is given by

$$[x_i, p_j] \equiv x_i p_j - p_j x_i = i\delta_{i,j}\hbar, \tag{2.2}$$

where $\hbar$ is Planck's constant $\hbar = \frac{h}{2\pi} = 1.05457173^{-34}$ m^{2}kgs^{-1}.

Importantly, a famous theorem due to Stone and von Neumann says that any implementation of the relation Eq. (2.2) among Hilbert space linear operators is unique up to *unitary equivalence*.[7] This means that, given any two ways of interpreting the relation Eq. (2.2) as a relation among linear operators on Hilbert spaces $\mathcal{H}_1$ and $\mathcal{H}_2$, there exists a one-to-one norm-preserving linear map $U : \mathcal{H}_1 \mapsto \mathcal{H}_2$ such that $UA_1U^{-1} = A_2$ for arbitrary linear operators defined on $\mathcal{H}_1$ and $\mathcal{H}_2$, respectively.

Due to the Stone–von Neumann theorem, the theories which historically were the first formulation of quantum mechanics – Schrödinger's wave mechanics and Göttingen matrix mechanics – can be seen as different Hilbert space representations of the same commutation relations (2.2). Owing to the unitary equivalence between them, choosing one of them rather than the other is a matter of computational convenience, not of physical significance. However, outside the scope of ordinary quantum mechanics with only finitely many degrees of freedom, e.g. in quantum statistical mechanics and quantum field theory where the number of degrees of freedom is infinite, unitarily *in*equivalent Hilbert space representations must be taken into account.[8] Many features of ordinary non-relativistic quantum mechanics are no longer valid in that context and the foundational challenges are even more complicated. Chapter 11 looks at some of the complications that arise.

If the Hilbert space $\mathcal{H}$ is finite-dimensional, any self-adjoint linear operator A can be decomposed as a sum of projection operators Π_i^A on subspaces of $\mathcal{H}$ in the form

$$A = \sum_i a_i \Pi_i^A. \tag{2.3}$$

The sum in Eq. (2.3) is called the *spectral decomposition* of A in terms of the projections Π_i^A. A generalised version, where the sum is replaced by an integral, holds in the infinite-dimensional case. As it turns out, in quantum theory the (necessarily real) numbers a_i are the possible values of A, i.e. the values A may be found to have upon measurement.

Let us now come to the mathematical expression of the quantum states, which, as remarked, are ascriptions of expectation values to observables. Mathematically, they are given by density matrices ρ, i.e. positive semi-definite self-adjoint linear operators on $\mathcal{H}$ whose trace equals one:

$$Tr(\rho) = 1. \tag{2.4}$$

The density matrix ρ ascribes an expectation value to the observable A according to the formula

$$\langle A \rangle_\rho = Tr(A\rho), \tag{2.5}$$

which provides the essential link between the mathematical formalism and its physical application.

Probabilities for the possible values of observables correspond to the expectation values of projection operators onto subspaces of $\mathcal{H}$. They are thus obtained from those instances of Eq. (2.5) where the place of A is taken by a projection operator Π_Δ^A that projects onto the span of eigenvectors of A with eigenvalues lying in the range of values Δ. Eq. (2.5) then becomes

$$\Pr_\rho(v(A) \in \Delta) = \langle \Pi_\Delta^A \rangle_\rho = Tr(\rho\, \Pi_\Delta^A), \tag{2.6}$$

where $v(A)$ denotes the value of A and Π_Δ^A. Eq. (2.6) is known as the *Born Rule*, honouring Born as the inventor of the 'statistical interpretation' of quantum theory. If the range Δ contains *no* eigenvalues of A, then, according to Eq. (2.6), $\Pr_\rho(v(A) \in \Delta) = 0$. In other words, the eigenvalues of the measured observables A are its only possible values. (Whether they are the only possible values *simpliciter* or merely *upon measurement* is an intricate question which can only be answered by settling on a particular *interpretation* of quantum theory. It will play an important role later in this work.)

There exists a lower bound on the product of standard deviations $\sigma_A = \sqrt{\langle A^2 \rangle - \langle A \rangle^2}$ and $\sigma_B = \sqrt{\langle B^2 \rangle - \langle B \rangle^2}$ of observables A and B in each quantum state ρ. Using the commutator $[A,B] = AB - BA$ it can be expressed as

$$\sigma_A \sigma_B \geq \frac{1}{2} \left| [A,B] \right| , \tag{2.7}$$

also known as the *Robertson uncertainty relation* (Robertson [1929]). The most famous instance of this relation is the *Heisenberg uncertainty relation* (Heisenberg [1927])

$$\sigma_{x_i} \sigma_{p_j} \geq \frac{\hbar}{2} , \tag{2.8}$$

which follows directly from Eq. (2.7) by inserting the canonical commutation relation Eq. (2.2) for $[A,B]$.

Of particular interest are those quantum states for which the associated density matrix has the property $\rho = \rho^2$ so that

$$Tr\left(\rho^2 \right) = Tr(\rho) = 1 . \tag{2.9}$$

Density matrices ρ having this property are referred to as *pure* quantum states, whereas those for which $Tr\left(\rho^2 \right) < 1$ are called *mixed* ones. Pure states correspond one-to-one to projection operators $\Pi_{|\psi\rangle}$. Using Dirac's so-called bra-ket notation, they are written as $|\psi\rangle\langle\psi|$, where the 'ket' $|\psi\rangle$, taken by itself, is simply the unit vector which spans $\Pi_{|\psi\rangle}$, and the 'bra' $\langle\psi|$ corresponds to the mapping from the vectors into the real numbers which takes $|\psi\rangle$ to 1 and vectors that are orthogonal to $|\psi\rangle$ to 0. Since $|\psi\rangle\langle\psi|$ projects onto the one-dimensional subspace spanned by $|\psi\rangle$, the pure states correspond one-to-one to one-dimensional subspaces (spanned by the vectors $|\psi\rangle$), which is why, when focusing only on pure quantum states, one often loosely speaking identifies quantum states with the Hilbert space *vectors* that span the one-dimensional subspaces.

A state vector $|\psi\rangle$ that is mapped by an observable A onto a multiple of itself is an *eigenstate* of A, i.e.

$$A|\psi\rangle = a|\psi\rangle \tag{2.10}$$

where the real number a is an eigenvalue of A. The eigenvalues a_i of an observable A are identical with the expansion coefficients in the spectral decomposition (2.3), whose coefficients in turn correspond to the possible values of A. State vectors that are *not* eigenstates of A can be written as sums of eigenstates of A. They are said to be *superpositions* of such eigenstates.

Schrödinger's wave mechanics corresponds to the Hilbert space representation on the space $\mathcal{L}^2(\mathbb{R}^n)$ of square-integrable complex-valued functions ψ over $\mathbb{R}^n$, for which

$$\int_{-\infty}^{\infty} \psi^*(x)\psi(x)d^n x < \infty. \tag{2.11}$$

When written as functions on this space, pure states are referred to as 'wave functions', and this notion is often used as a synonym with 'pure state'. Since the space $\mathbb{R}^n$ on which wave functions are defined is generally not the familiar three- (or four-) dimensional *physical* space, the function ψ cannot be straightforwardly interpreted as a 'field' on the space-time manifold. Acknowledging this, however, has no direct ramifications for whether or not ψ is a physically *real* quantity. The question of whether it is will take centre stage in this work.

In non-relativistic quantum mechanics, the time-dimension has a different status from the space-dimensions in that it is represented by a *parameter* and is not treated as an observable in its own right. This is reflected in the fact that it is often more convenient to conceive of time-evolution as a feature of quantum states rather than observables. The formalism where the quantum states are treated as time-dependent is called the *Schrödinger picture*. If we denote by 'H' the 'Hamiltonian' operator, which corresponds to the energy observable, the time-evolution of the quantum state ρ follows the von Neumann equation

$$i\hbar\frac{\partial\rho}{\partial t} = [H,\rho]. \tag{2.12}$$

For pure states, i.e. state vectors $|\psi\rangle$, this reduces to the Schrödinger equation

$$i\hbar\frac{\partial}{\partial t}|\psi\rangle = H|\psi\rangle. \tag{2.13}$$

It is equally possible to treat the quantum states as time-dependent and to cover any time dependences by the observables. The resulting formalism is called the *Heisenberg picture*. For observables $A(t)$ without any *explicit* dependence on time (i.e. observables for which $\partial_t A(t)=0$), their time-evolution is governed by the Heisenberg equation, which reads

$$\frac{d}{dt}A(t) = \frac{i}{\hbar}[H,A(t)]. \tag{2.14}$$

Given the Hamiltonian H of a quantum system, the formalism just outlined allows one to make testable predictions about the values of observables by computing probabilities from the quantum states using Eq. (2.6). In the next section, I discuss how the textbook account of quantum theory treats situations when some observable of a quantum system is *measured*.

2.2　Measurement and collapse

Quantum states, as explained in the previous section, ascribe probabilities to the possible values of observables. They do not normally tell us anything about their *actual* values. If we assume that quantum theory is a complete and fundamental theory of the world that describes *all* properties of quantum systems, it seems natural to conclude that only those values of observables to which probability 1 is ascribed by the quantum state of the system are physically real. And indeed, in accordance with this line of thought, the orthodox answer to the question of which observables of a quantum system have determinate values is the *eigenstate/eigenvalue link*,[9] which claims that

$$v_{|\psi\rangle}(A) = a, \quad \text{iff} \quad A|\psi\rangle = a|\psi\rangle, \tag{2.15}$$

or, in words, that an observable A of a system in the quantum state $|\psi\rangle$ has a determinate value $v_{|\psi\rangle}(A) = a$ if and only if that state $|\psi\rangle$ is an eigenstate of (the self-adjoint linear operator associated with) A to the eigenvalue a. For, by the Born Rule Eq. (2.6), in that case $\Pr_{|\psi\rangle}(v(A) = a) = 1$. Conversely, given some observable A and a quantum state $|\psi\rangle$, unless Eq. (2.15) holds for some a, according to the eigenstate/eigenvalue link A does not have a determinate value in $|\psi\rangle$ at all.

The main problem with the eigenstate/eigenvalue link is its consequence that almost no observables of a quantum system have determinate values. This seems particularly unacceptable with respect to experimental contexts where a value a is obtained that is *not* an eigenvalue of the measured observable A. For, as we have seen, the eigenstate/eigenvalue link implies that unless $|\psi\rangle$ happens to be an eigenstate of the measured observable A and therefore ascribes probability 1 to exactly one possible value a of A and probability 0 to all others, none of the possible values of A is realised. It seems unclear why, given this assumption, measuring A should result in any determinate measurement outcome at all. This difficulty makes it overwhelmingly natural to question the eigenstate/eigenvalue link (2.15) and consider replacing it with a more permissive criterion of in which conditions observables have determinate values.

The challenge of accounting for determinate measurement outcomes is a version of the *measurement problem* mentioned already in Chapter 1. It is an extremely difficult problem due to the fact that it is completely unclear how Eq. (2.15) might be relaxed in a natural way. There are various no-go results on the assignment of determinate values of observables (the Kochen–Specker theorem in particular; see Appendix B) which show that the task of relaxing Eq. (2.15) does not have any straightforward solution.

Textbook accounts of quantum mechanics avoid the measurement problem by invoking what is widely seen as an act of brute force, namely, by resorting to the notorious *collapse of the wave function*. This prescription for dealing with quantum systems being measured was originally introduced by Heisenberg in his seminal paper (Heisenberg [1927]) and later made into an official part of quantum theory in von Neumann's mathematical codification (von Neumann [1955]). According to the collapse postulate, the state of a quantum system that is measured is projected on the linear span of eigenstates to the measured observable that are compatible with the measured result. Formally, for a quantum system in the quantum state $|\psi\rangle$, if the measured result lies in the range Δ, the collapse prescription is

$$|\psi\rangle \mapsto \frac{\Pi_\Delta |\psi\rangle}{|\Pi_\Delta |\psi\rangle|}, \tag{2.16}$$

where Π_Δ denotes the projection onto the linear span of eigenvectors of A with eigenvalues lying within Δ. If the Hilbert space subspace of eigenvectors of the measurement observable A to the eigenvalue a_i (the measured result) is one-dimensional, this reduces to

$$|\psi\rangle \to |a_i\rangle, \tag{2.17}$$

where $A|a_i\rangle = a_i|a_i\rangle$.

Using measurement collapse, the problem of accounting for determinate values of measured observables is avoided inasmuch as the post-measurement state is set to $|a_i\rangle$ by *fiat* in virtue of Eq. (2.16). For this latter state, the eigenstate/eigenvalue link tells us that A now does have the value a_i, in conformity with what we believe to know about the system as a consequence of the measurement. Quantum probabilities as determined by the Born Rule (2.6) are interpreted here as probabilities of *collapse* into any of the possible post-measurement states $|a_i\rangle$. This is how measurement collapse overcomes the problem of determinate values as far as practical purposes are concerned.

However, measurement collapse has never been an uncontested part of quantum theory. There are excellent reasons to worry about it. Laura Ruetsche expresses the widespread concerns about it as follows:

> Recognizing this [measurement] problem, von Neumann [...] responded by invoking the *deus ex machina* of measurement collapse, a sudden, irreversible, discontinuous change of the state of the measured system to an eigenstate of the observable measured. [...] Collapse is a Humean miracle, a violation of the law of nature expressed by the Schrödinger equation. If collapse and unitary evolution are to coexist in a single, consistent theory, situations subject to unitary evolution must be sharply and unambiguously distinguished from situations subject to collapse. [...] [D]espite evocative appeals to such factors as the intrusion of consciousness or the necessarily macroscopic nature of the measuring apparatus, no one has managed to distinguish these situations clearly. (Ruetsche [2002], p. 209)

The two main worries expressed in this passage are the following: first, that in contrast to the smooth time-evolution governed by the Schrödinger equation, collapse is sudden and discontinuous; second, that we are not given any clear criterion for distinguishing between situations where collapse occurs and situations where it does not. The first of these worries, concerning the sudden and instantaneous character of collapse, arises because it clashes with the principle of relativity theory that there are no preferred planes (or surfaces) of simultaneity on which collapse could naturally be taken to be instantaneous. I elaborate on this problem in the following section when discussing the quantum mechanical treatment of many-component systems, and again in Section 4.4.

The second worry concerning collapse – the complaint that we are not given any precise criterion of under which conditions it occurs – is perhaps even more disturbing. Usually, it is said that collapse occurs whenever the system is measured, but this raises the immediate concern that anthropocentric notions such as 'measurement' and 'measurer' are too imprecise and not sufficiently fundamental to account for what happens under which conditions in a satisfactory way. Highlighting this problematic aspect of collapse, John S. Bell sarcastically asks:

What exactly qualifies some physical systems to play the role of 'measurer'? Was the wavefunction of the world waiting to jump for thousands of millions of years until a single-celled living creature appeared? Or did it have to wait a little longer, for some better qualified system [...] with a PhD? If the theory is to apply to anything but highly idealised laboratory operations, are we not obliged to admit that more or less 'measurement-like' processes are going on more or less all the time, more or less everywhere? Do we not have jumping then all the time? (Bell [2007], p. 34)

It seems hard not to agree with Bell here that the standard account, according to which collapse occurs whenever the system is measured, is completely unsatisfying by placing a conceptual burden on anthropocentric notions such as 'measurement' that they cannot possibly bear. Nevertheless, collapse remains an integral part of quantum theoretical practice, and all suggested interpretations have to account for why it works so well in that practice.

In the following section I review the quantum mechanical treatment of multi-component quantum systems and the notion of quantum entanglement.

2.3 Many-component systems and entanglement

Quantum systems that are not legitimately considered as isolated from each other must be treated as jointly forming a *multi-component* quantum system. The Hilbert space associated with a multi-component system is the *tensor product* space $\mathcal{H} = \mathcal{H}_1 \otimes \ldots \otimes \mathcal{H}_N$, where the Hilbert spaces associated with the individual systems are $\mathcal{H}_1, \ldots, \mathcal{H}_N$. The properties of the tensor-product space $\mathcal{H}_1 \otimes \ldots \otimes \mathcal{H}_N$ are very different from those of the Cartesian-product space $\mathcal{H}_1 \times \ldots \times \mathcal{H}_N$. For example, the dimension of the tensor-product space $\mathcal{H}$ equals the *product* of the dimensions of the individual Hilbert spaces $\mathcal{H}_1, \ldots, \mathcal{H}_N$, whereas the dimension of the Cartesian-product space $\mathcal{H}_1 \times \ldots \times \mathcal{H}_N$ equals the *sum* of the dimensions of the spaces $\mathcal{H}_1, \ldots, \mathcal{H}_N$. On the tensor-product space $\mathcal{H}$, the scalar product $\langle \ldots | \ldots \rangle$ is defined in terms of the scalar products $\langle \ldots | \ldots \rangle_1, \ldots, \langle \ldots | \ldots \rangle_N$ of the individual Hilbert spaces $\mathcal{H}_1, \ldots, \mathcal{H}_N$ by

$$\langle \phi_1 \otimes \ldots \otimes \phi_N | \psi_1 \otimes \ldots \otimes \psi_N \rangle = \langle \phi_1 | \psi_1 \rangle_1 \ldots \langle \phi_N | \psi_N \rangle_N \tag{2.18}$$

for arbitrary $|\phi_1\rangle, |\psi_1\rangle \in H_1, \ldots, |\phi_N\rangle, |\psi_N\rangle \in H_N$.

For many-component quantum systems that are composed of quantum systems of the same *type* of system (such as electron, photon, charm-quark, etc.), the appropriate Hilbert space to consider is not the full $\mathcal{H} = \mathcal{H}_1 \otimes \dots \otimes \mathcal{H}_N$, but one of its subspaces, namely, either the one which consists of those vectors in $\mathcal{H}$ that are fully *symmetric* under exchange among labels $1, \dots, N$ or the one which consists of those vectors that are fully *anti-symmetric* under such exchanges. In the symmetric case, switching two system labels does not change the state vector; in the anti-asymmetric case, switching two system labels is the same as applying an overall minus sign.

Particles for which the associated Hilbert space consists of the symmetric vectors are called *bosons*, and particles for which it consists of the antisymmetric ones are called *fermions*. The fact that fermionic state vectors must be permutation anti-symmetric entails that no two fermions can be in the same quantum state, which means that they cannot agree in all their '*quantum numbers*': the expansion coefficients that specify their quantum states in terms of a single-system Hilbert space basis. This is Pauli's famous *exclusion principle*; a statement which plays an essential role in many explanations and predictions of quantum theory. For example, it allows an immediate understanding of some of the most elementary features of the periodic table of elements.

The state vector of some many-component quantum system in the tensor product Hilbert space $\mathcal{H} = \mathcal{H}_1 \otimes \dots \otimes \mathcal{H}_N$ can in general not itself be written in product form as $|\psi\rangle = |\psi_1\rangle \otimes \dots \otimes |\psi_N\rangle$ with $|\psi_1\rangle \in H_1$, $\dots$, $|\psi_N\rangle \in H_N$. If it cannot be written in product form, it is called an *entangled* quantum state. Entangled quantum states give rise to some of the most surprising and spectacular predictions made by quantum theory. They are exploited, in particular, in the flourishing discipline of *quantum information theory*. The term 'entanglement' itself was originally introduced by Schrödinger in 1935 in a paper, where he characterises the phenomenon as follows:

> When two systems, of which we know the states by their respective representatives, enter into temporary physical interaction due to known forces between them, and when after a time of mutual influence the systems separate again, then they can no longer be described in the same way as before, viz. by endowing each of them with a representative of its own. I would not call that *one* but rather *the* characteristic trait of quantum mechanics, the one that enforces its entire departure from classical lines of thought. By the interaction the two representatives (or ψ-functions) have become *entangled*. To

disentangle them we must gather further information by experiment, although we knew as much as anybody could possibly know about all that happened. (Schrödinger [1935], p. 555, emphasis of *'one'* and *'the'* due to Schrödinger, further emphases by me)

As Schrödinger emphasises, entanglement is dissolved only when one of the subsystems of the joint system is measured. Formally, the 'disentanglement' in this case is a consequence of the application of measurement collapse to the state of the measured system. Since this application of collapse affects the states of the other (sub-) systems independently of the spatial distance that lies between the (sub-) systems, entanglement aggravates the above-mentioned tension between collapse and the requirements of special relativity.

Based on considerations on entangled states and an assumption of *locality*, Einstein, Podolsky, and Rosen (EPR) argue in a famous paper (Einstein et al. [1935]) that quantum states cannot possibly give *complete* descriptions of quantum systems. The clearest version of the argument Einstein seems to have had in mind when contributing to it does not appear in the paper itself but in his personal correspondence and his later writings.[10] It is most conveniently given in terms of a two-system setup, where the two individual systems are prepared in such a way that an entangled state must be assigned to the combined system. One of the simplest such setups consists of two systems of which only the so-called *spin* degrees of freedom are considered, which, for each system, are associated with a Hilbert space of complex dimension 2 (isomorphic to $\mathbb{C}^2$). The observables corresponding to the spatial directions of spin are denoted by 'S_x', 'S_y', and 'S_z', and the possible measured values are for all directions $+1/2$ and $-1/2$. One possible entangled quantum state of the combined system (conventionally labelled 'EPRB' for 'EPR-Bohm') predicts perfect anti-correlations for measurements of the same component S_i at each system:

$$|\psi_{EPRB}\rangle = \frac{1}{\sqrt{2}}\left(|+\rangle_1|-\rangle_2 - |-\rangle_1|+\rangle_2\right), \tag{2.19}$$

where $|+\rangle$ ($|-\rangle$) is the eigenstate with eigenvalue $+1/2$ ($-1/2$) to the spin observable in, say, the z-direction. The quantum state $|\psi_{EPRB}\rangle$ patently cannot be written as a product of state vectors from the Hilbert spaces $\mathcal{H}_1$ and $\mathcal{H}_2$ associated with the individual subsystems.

With respect to this setup one considers an agent Alice, located at the first system and performing a measurement of, say, S_z. Having registered the result, Alice applies the collapse postulate and afterwards assigns two distinct and no longer entangled states to the two systems, which in general depend both on the choice of observable measured and on the measured result. For example, if she registers the outcome $+1/2$, the projection postulate commits her to assigning the post-measurement state $|+\rangle_1$ to her own and the post-measurement state $|-\rangle_2$ to Bob's system. As a consequence, the state she assigns to the second system after measurement will in part depend on her choice of direction of spin measured at her own system, even though the two systems can be located as far apart as one might wish. Here EPR's assumption of locality comes into play, and it dictates in this case that Alice's measurement cannot instantaneously influence any physical properties of the second, distant, system. If we accept the locality assumption, this means that the physical properties of the second system in the moment of measurement and immediately after it are expressed by different quantum states, depending on what choice of measurement of her own system Alice has made. If, conversely, we assume that quantum states correspond one-to-one to the physical properties of quantum systems, the physical influences involved must travel arbitrarily fast.

A natural response to this difficulty is to consider modifying the dynamics of collapse, such that it is no longer instantaneous but rather propagates with the velocity of light, that is, in the language of special relativity, along light-like hypersurfaces. The difficulty with this suggestion is that, for measurement events at space-like separation from each other, such a retarded version of collapse would mean that both measurements are carried out on the uncollapsed state, which would seem to leave the persistent correlation between such results unaccounted for. In the words of Maudlin, '[s]ince polarization measurements can be made at space-like separation, and since the results on one side must be influenced by the collapse initiated at the other, delayed collapses won't work' (Maudlin [2011], p. 180).

The problem becomes especially dramatic in situations where measurements are performed at both systems by different agents in such a way that the distance between the measurement events (or processes) is space-like, perhaps even in such a way that each measurement is carried out first in its own inertial rest frame. Such a setup can indeed be experimentally realised, as demonstrated by (Zbinden et al. [2001]), whose experimental results confirm the predictions of standard quantum theory. In this case there is no non-arbitrary answer at all to the question of which measurement occurs first, such that the measurement collapse associated with it might trigger the abrupt change in the quantum state of the other. The problem of reconciling collapse with relativity seems especially troublesome here, if not hopeless.

Non-locality is revisited in more detail in later chapters (in particular Section 4.4 and Chapter 10). In a next step, I review the main interpretations of quantum theory and how they react to the measurement problem.

3
Interpretations as Solutions to the Measurement Problem

Interpretations of quantum theory can be usefully categorised according to which moves they make to avoid the measurement problem. The chapter gives a more systematic exposition of that problem, followed by an overview of the answers that are given by the most popular interpretations. As we shall see, it is helpful to group the interpretations together according to which of the assumptions that lead to the measurement problem they reject.

3.1 Exposition of the problem

3.1.1 Maudlin's formulation

There are various different formulations of the measurement problem in the literature, which are sometimes even characterised as different measurement problems. In what follows I focus on an exposition due to Maudlin, who considers three different formulations.[11] I shall focus mainly on the first. (The second is a challenge to the Everett interpretation and is going to be brought up later in this chapter in Section 3.4.2. The third is not directly relevant to our present purposes.) In none of these formulations does the problem rest on the controversial eigenstate/eigenvalue link Eq. (2.15), which indicates that the problem is of a very general nature.

Maudlin's first measurement problem – the one which will chiefly concern us here – has the form of an incompatibility between the following three plausible-looking assumptions:

1.A The wave-function of a system is complete, i.e. the wave function specifies (directly or indirectly) all of the physical properties of a system.

1.B The wave-function always evolves in accord with a linear dynamical equation (e.g. the Schrödinger equation).

1.C Measurements of, e.g., the spin of an electron always (or at least usually) have determinate outcomes, i.e., at the end of the measurement the measuring device is either in a state which indicates spin up (and not down) or spin down (and not up). (Maudlin [1995], p. 7)

To see that these assumptions are incompatible, consider a situation in which the system being measured is not in an eigenstate of the measured observable. In that case, according to 1.B, the state of the combined system, consisting of the measured system together with the measuring apparatus, evolves into a superposition of (products of) eigenstates of the measured observable and the 'pointer' observable. The latter is defined by the fact that its different possible values correspond to the macroscopically different pointer or display configurations of the apparatus. This state does not single out or prefer by itself any of the possible values of the measured observable of the measured system nor of the pointer observable pertaining to the apparatus. Nevertheless, according to 1.A, it provides a complete description of the combined system, including both the measured system and the apparatus. So, we must conclude that none of the possible values of the measured and the pointer observables is actually realised (or that all of them are, as some radical interpreters conclude). Assumption 1.C, however, requires that at the end of the measurement process the value of the pointer observable be determinate.

To conclude, the three assumptions, taken together, are incompatible and at least one of them has to be dropped. As a consequence, a necessary condition for an interpretation of quantum theory to count as a candidate *solution* to the measurement problem is to declare either 1.A or 1.B or 1.C as wrong (or to find a loophole in the reasoning that establishes their incompatibility, which, however, seems very difficult).

Additional light can be cast on the problem by formulating it in terms of a model of ideal quantum measurement: in this model, prior to measurement the measured system and the apparatus are in pure states $|\psi\rangle$ and $|\phi\rangle$ (the assumption of purity is inessential and can easily be dropped), which means that the resulting state of the combined system is $|\Psi_{pre}\rangle = |\psi\rangle \otimes |\phi\rangle$.

Next, one considers an eigenbasis $\{|s_i\rangle\}$ to the observable of the measured system which is taken to be measured and an eigenbasis $\{|a_i\rangle\}$ to the pointer observable of the apparatus, so that the complete pre-measurement state takes the form

$$|\Psi_{pre}\rangle = |\psi\rangle \otimes |\phi\rangle$$

$$= \sum_i c_i |s_i\rangle \otimes \sum_i d_i |a_i\rangle \tag{3.1}$$

$$= \sum_{i,j} c_i d_j |s_i\rangle \otimes |a_j\rangle$$

with suitable coefficients c_i and d_i.
Measurement involves an interaction between the measured system and the apparatus. The better the apparatus is suited to serving as a *measurement* apparatus, the better will be the resulting correlation between measured system observable eigenstates $|s_i\rangle$ and the pointer eigenstates $|a_i\rangle$ in the post-measurement state. Thus, ideally, the measurement interaction leads to a post-measurement state that can be written as

$$|\Psi_{post}\rangle = \sum_i c_i' |s_i\rangle \otimes |a_i\rangle \tag{3.2}$$

with the same coefficients $c_i' = c_i$ for the measured system state that were also used in the expansion of $|\psi\rangle$ in the basis $\{|s_i\rangle\}$. If the measurement is less than ideal, then either $c_i' \neq c_i$ or $|\Psi_{post}\rangle$ contains contributions from terms of the form $|s_i\rangle \otimes |a_j\rangle$ with $i \neq j$. However, the presence or absence of such terms has no impact on the presence of the problem: it arises from the crucial difficulty that $|\Psi_{post}\rangle$ as the post-measurement state does not in any way single out (or *prefer*) any of the measured system states $|s_i\rangle$ or apparatus states $|a_i\rangle$. In particular, none of them are revealed in any way as the *real* states of the system and apparatus, and, correspondingly, none of the associated eigenvalues s_i and a_i are suggested as the actual values of observables at the outset of the measurement process. This result is, as it stands, incompatible with the manifest empirical fact that measurements actually *do* have determinate outcomes. It is evidently not of any help to allow that the measurement interaction may be *less* than ideal, for allowing that c_i' may to some degree deviate from c_i or including contributions proportional to $|a_i\rangle \otimes |s_j\rangle$ with $i \neq j$ does not fundamentally change the situation.

To conclude, unless one is prepared to deny either assumption 1.A – that the quantum state gives a complete description of the quantum system – or assumption 1.B that there exist no further dynamical principles of state-evolution besides the Schrödinger equation, one encounters a blatant conflict with an incontestable feature of our scientific and everyday lives: that measurements have determinate outcomes.

Since the problem arises for arbitrary situations where the macroscopic properties of a physical system are assumed to be determinate and not only for 'measurement' situations in the narrow sense of the word, it is also (and appropriately so) referred to as the '*macro-objectification problem*' (Ghirardi [2011]).

3.1.2 A fallacious solution

Sometimes, a fallacious solution to the measurement problem is given, the core idea of which I will briefly discuss. It starts by considering the density matrix associated with the state (3.2), which (assuming an ideal measurement with $c_i' = c_i$ for all i) is given by

$$\rho_{\text{post}} = |\Psi_{\text{post}}\rangle\langle\Psi_{\text{post}}| = \sum_{i,j} c_i c_j^* |s_i\, a_i\rangle\langle s_j\, a_j| \tag{3.3}$$

(where $|s_i\, a_i\rangle$ is a shorthand for $|s_i\rangle \otimes |a_i\rangle$) and supposedly describes the combined system in the aftermath of ideal measurement. Next, one considers what seems to be the natural candidate if one wants to ascribe some quantum state to the measured system *alone*, i.e. independently of its being part of a larger multi-component system together with the apparatus. This is the so-called *reduced* density matrix $\rho_{\text{post,S}}$, which is obtained from ρ_{post} by taking the trace over the apparatus degrees of freedom. It is given by

$$\rho_{\text{post,S}} = \sum_i |c_i|^2 |s_i\rangle\langle s_i|. \tag{3.4}$$

This density matrix, taken by itself, corresponds to a mixed state, and the (fallacious) argument goes that it expresses our subjective ignorance as regards which of the pure states $|s_i\rangle$ is actually realised. In this interpretation, the coefficients $|c_i|^2$, corresponding to the Born Rule probabilities, are expressions of our subjective ignorance as to which of the possible outcomes occurs. Since the same argument can be applied to the reduced density matrix of the apparatus, the fallacious reasoning goes, this resolves the measurement problem.

Sometimes, an analogous argument is given for the combined system, considered jointly with the environment in which it is placed. Here it is typically noted that realistic interactions between macroscopic systems such as measurement apparatuses and their environment lead very quickly to entangled states, which are, approximately, of the form

$$|\Psi'_{\text{post}}\rangle = \sum_i c_i |s_i\rangle \otimes |a_i\rangle \otimes |e_i\rangle, \tag{3.5}$$

where the states $|e_i\rangle$ form a basis of environment states. As the comparison between Eqs. (3.2) and (3.5) shows, the interaction with the environment has an effect analogous to an ideal measurement (with 'ideal' in the same sense as above) of the system that consists of the measured system together with the apparatus by the environment.

The reduced density matrix for the measured system and the apparatus, taken together, is now given by

$$\rho'_{\text{post,SA}} = \sum_i |c_i|^2 |s_i a_i\rangle\langle s_i a_i|. \tag{3.6}$$

Formally, this is precisely the desired result, because $\rho'_{\text{post,SA}}$, taken by itself, is interpretable as expressing ignorance concerning which of the states $|s_i a_i\rangle$ is realised, each one with a probabilistic weight of $|c_i|^2$. The mechanism that leads to a reduced state of the form (3.6) is known as 'environment-induced decoherence' (or simply 'decoherence'), and the argument just outlined is sometimes taken to establish that decoherence by itself *solves* the measurement problem.

To see why this argument fails – at least as it stands – one must distinguish between 'properly' and 'improperly' mixed states – terms that are due to (d'Espagnat [1976]). The 'proper mixtures' are assigned to quantum systems in cases where the state preparation procedure does not narrow down possible states to assign to a unique pure state, or where systems arriving from different pure state preparation devices are inextricably intermingled. It is only in these cases where the ignorance interpretation of mixed state makes sense in that it claims that the assignment of a mixed state reflects the observer's ignorance about the pure state the system really is in. Contrasting with the 'properly mixed states' are the 'improper mixtures'. These are mixed states that one obtains for the subsystems of many-component systems by performing the trace operation over the degrees of freedom of the other subsystems and arriving at a formally mixed state, exactly as above for $\rho_{\text{post,S}}$ and $\rho'_{\text{post,SA}}$.

Improper mixtures cannot be given an ignorance interpretation in the sense of reflecting ignorance about the 'true' pure state of the quantum system. This is due to the fact that the state ρ of the combined (multi-component) system does not have the form of a mixture of states which are products of pure states for the subsystems in such a way that the reduced density matrices of these subsystems are mixtures of these pure states with the coefficients used in the decomposition of the state ρ of the combined system. Consequently, the reduced density matrices of the subsystems cannot be given an interpretation as reflecting ignorance about any pure states the subsystems are in fact in, so the ignorance interpretation of mixed states applies only to proper, not to improper, mixtures.

To conclude, the incompatibility of assumptions 1.A–1.C must be taken seriously. To avoid it, one must reject at least one of Maudlin's assumptions 1.A, 1.B, and 1.C. The following section of this chapter investigates the prospects for *solutions* of the problem: accounts that reject at least one of 1.A, 1.B, and 1.C as *wrong*. The prospects for *dissolutions* – accounts that reject at least one of 1.A, 1.B, and 1.C for being based on *mistaken conceptual presuppositions* – are explored in the following chapter.

Giving a complete overview of the suggested solutions to the measurement problem is besides the scope of a single chapter of this work. However, a brief recapitulation of exemplary and especially well-known interpretations and their virtues and drawbacks will be useful for providing a backdrop against which the virtues and drawbacks of the account developed in later chapters of this book can be seen. Those solutions to the measurement problem which reject assumption 1.A are reviewed in Section 3.2, those which reject 1.B in Section 3.3, and those which reject 1.C in Section 3.4.

3.2 Additional parameters

Interpretations that regard the quantum state as an *incomplete* description of a physical system are often referred to (often somewhat misleadingly) as 'hidden variable' interpretations. Independently of how the additional variables are introduced, interpretations which postulate them must respect various no-go results, most importantly the theorems due to Bell and Kochen–Specker, see the brief reviews in Appendices A and B. Bell's theorem is commonly interpreted as implying that any hidden variable interpretation of quantum theory must in some sense be 'non-local' (But this terminology can arguably be misleading; see Chapter 10). The Kochen–Specker theorem states that hypothetical complete assignments of determinate values to observables must be *contextual* by breaking the usual one-to-one correspondence between observables and self-adjoint linear operators. Pilot wave theory, perhaps the paradigm approach that rejects assumption 1.A, nicely exemplifies ways of how the threat of running into these theorems can be avoided.

3.2.1 Pilot wave theory

Pilot wave theory is probably the best worked-out theory that uses additional parameters to complement quantum states in the description of quantum systems. It is also known as the de Broglie/Bohm theory (and as Bohmian mechanics), named after its (independent) inventors Louis de Broglie and David Bohm. The additional parameters which it postulates in addition to the wave function of a quantum system are determinate particle positions, whose time-evolution is assumed to be governed by the influence of the wave function. Since the wave function acts on the particle configuration, but not vice versa, the wave function is often referred to as the *guiding wave* (or *pilot wave*, hence the name of the theory).

The formalism

In pilot wave theory, the time-evolution of the wave function remains governed by the Schrödinger equation. The law of time-evolution for the particle positions, which incorporates the nature of the wave function as a guiding wave, is called the *guiding* equation. This equation says that the time-derivative of the *j*-th coordinate of a quantum system is proportional to the ratio between the quantum probability current $J_j = \mathrm{Im}\left(\psi^*\partial_j\psi\right)$ and the quantum probability density $\rho = \psi^*\psi$. Formally, for a system with n degrees of freedom, where q_j is the value of the *j*-th position coordinate and m_j the mass associated with the *j*-th degree of freedom, the time-derivative of q_j (i.e. the particle velocity) is given by

$$m_j \frac{dq_j(t)}{dt} = \hbar\,\mathrm{Im}\left(\frac{\psi^*\partial_j\psi}{\psi^*\psi}(q_1,q_2,\ldots,q_nt)\right). \tag{3.7}$$

As repeatedly emphasised by its proponents, pilot wave theory is only misleadingly called a 'hidden variables theory': the particle configurations themselves are not *hidden* at all; they are what we have rather direct empirical access to, not the wave function that guides their evolution in time. As noted by Bell, '[i]n any case, the most hidden of all variables, in the pilot wave picture, is the wavefunction, which manifests itself only by its influence on the complementary variables' (Bell [2004], p. 202). However, the wave function remains an objective physical quantity in pilot wave theory in that there is exactly one correct wave function for the whole world, governing the time-evolution of the particles in it in accordance with the guiding equation Eq. (3.7).

In order to recover the usual probabilistic predictions of quantum theory in pilot wave theory, one must make an assumption as regards the actual distribution of particles in our world. Due to the specific features of the guiding equation, where the numerator of the right-hand side equals the quantum probability current and the denominator the quantum probability density, it follows from the (easily provable) quantum probability continuity equation $\partial_t \rho + \text{div} J = 0$ (with $\rho = \psi^* \psi$ and $J = \text{Im}\left(\psi^* \nabla \psi\right)$) that the distribution remains stable as soon as it equals ρ. For that reason, the condition that the distribution of particle configurations in the universe conforms to $\psi^* \psi$ is often referred to as *'quantum equilibrium'*. In contrast to ordinary quantum theory, the quantity $\psi^* \psi$ is only *contingently* equal to the configuration distribution in the sense that there may in principle be conditions where the latter is substantially different from $\psi^* \psi$.

Measurement in pilot wave theory

Pilot wave theory differs radically from standard quantum theory in its account of quantum measurement. The main idea of the pilot wave approach to measurement is that what actually happens in 'measurement' situations is that some pointer position (or display configuration) is registered. Pilot wave theory construes all 'measurements' as ultimately determining the position of whatever plays the role of the 'pointer' in the experiment. 'Measurement', thus construed, does not usually help us establish the *actual*, pre-existing values of those observables that we normally take to be 'measured'. For example, on the pilot wave account of the hydrogen atom, the electron is at permanent rest in the hydrogen ground state. If, however, we apply the Born Rule to compute the expectation value of its velocity in this state, we obtain a nonzero value. The apparent conflict is resolved by noting that, from the point of view of pilot wave theory, the alleged 'measurement' of the electron velocity (inasmuch as we can actually carry it out) is in fact an experiment resulting in a determinate pointer position for which we can make probabilistic predictions. It does not in general reveal the actual value of the observable associated with the operator used to compute the outcome probability.

To conclude, from the point of view of pilot wave theory, the connection between observables and operators is helpful for predicting the outcomes of experiments, but it is misleading to think of it as obtaining in a more profound sense over and above this operational one. Rejecting any such 'naive' one-to-one association of observables with Hilbert space operators allows pilot wave theory to embrace contextuality (in the sense of the Kochen–Specker theorem) as a natural feature. Since the theorem can only be derived by making an assumption of *non*-contextuality, which, as just seen, pilot wave theory denies, pilot wave theory is not in any way threatened by it.

The account of measurement in pilot wave theory does not stop at accepting contextuality, however, but goes on to recover the collapse of the wave function as an *effective* concept: it explains why applying collapse to the quantum state when making measurements works as well as it does without at the same time acknowledging it as an independent law of the time-evolution of wave functions in its own right. To describe measurement situations, one has to take into account the measuring system and its particle configuration and consider the wave function of the complete (system plus apparatus) system. If we denote this wave function (for the complete system) by Ψ, the so-called *conditional* wave function for the measured system is given (in the position representation) by $\psi(x) = \Psi(x, Q_A)$, where Q_A denotes the actual apparatus configuration.

The time-evolution of the position degrees of freedom associated with the measured system is given by inserting the *conditional* wave function $\psi(x)$ on the right-hand side of the guiding equation Eq. (3.7). If the system is sufficiently well isolated from its environment (including the apparatus, if there is one), the time-evolution of the conditional wave function itself is governed by the Schrödinger equation for the system degrees of freedom alone. This explains why neglecting the environment of a sufficiently isolated quantum system works so well in ordinary quantum mechanics. It can further be shown that, whenever the condition of quantum equilibrium is fulfilled (in that the overall particle distribution in the world corresponds to $\psi^*\psi$), the conditional wave function is subject to instantaneous 'collapses' of precisely the form required in the standard formalism when measurements are performed. Contrary to the standard formalism, however, pilot wave theory can account precisely and without invoking any anthropocentric vocabulary for under which conditions (and to which degree of numerical approximation) the *effective* collapses which it countenances do occur.

A final feature of pilot wave theory that deserves to be mentioned is its manifestly non-local character: in accordance with the guiding equation, the velocity of a particle will typically depend on the positions of all other particles *at the same time*, however distant in space they may be. As a consequence of this appeal to an absolute simultaneity relation, the question of how pilot wave theory may be reconciled with the requirements of *relativistic* space-time is an intricate one. The majority view among proponents of pilot wave theory (such as (Bohm and Hiley [1993]), (Holland [1993]), (Valentini [1997])) seems to be that Lorentz invariance is to be recovered as an *emergent* symmetry of our observations, which is absent at a more fundamental level. Advocates of pilot wave theory typically hold that the manifest non-locality of their approach is a *virtue* rather than a drawback, as it makes explicit the supposedly unavoidable non-locality that any theory must have which recovers the predictions of quantum theory, rather than sweeping it under the rug as the standard formalism allegedly does. As I shall argue, however, in Chapter 10, there is no genuine 'non-locality' (worth the name) involved in quantum theory, which means that we should perhaps not welcome the explicit non-locality pilot wave theory introduces.

Objections and challenges

Pilot wave theory has been criticised in many respects, often unfairly, based on spurious grounds and misunderstandings. Sometimes it is rejected on purely aesthetic grounds, for example by Rudolf Peierls, declaring that 'I must confess that the scheme, with both hidden variables and probability rules seems to me exceedingly ugly' (Peierls [1991], p. 20). However, the rejection of pilot wave theory on aesthetic grounds is problematic, as convincingly emphasised by Putnam, who argues that, as an example, '[t]he formula for the velocity field is extremely simple: you have the probability current in the theory anyway, and you take the velocity vector to be proportional to the current. There is nothing particularly inelegant about that; if anything, it is remarkably elegant!' (Putnam [2005], p. 622). Even though this is largely an aesthetic question, it seems hard to deny that Putnam has a point here.

Another criticism which has been brought forward against pilot wave theory is that it *unnecessarily* postulates determinate particle configurations, which, as the supposed viability of the Everett interpretation shows (that will be discussed in Section 3.4), are not actually needed to avoid the shortcomings of the standard textbook account laid out in Chapter 2. The idea here is that pilot wave theory postulates that the universal wave function is non-vanishing even in those (enormous) parts of configuration space that are far from the actual particle configuration, which seems to mean that it makes the same assumptions concerning quantum states as the Everett interpretation. This objection has been made, for instance, by David Deutsch, who derides versions of pilot wave theory for being 'parallel-universe theories in a state of chronic denial' (Deutsch [1996], p. 225). From the standpoint of pilot wave theory itself, this objection is puzzling: the determinate particle configurations are not surplus structure that may be left outside the picture. Rather, the entire envisaged solution of the measurement problem rests on their existence, while the Everett interpretation has no similar elegant answer.

Both objections just mentioned are arguably flawed, but they may have a justified core. Perhaps most physicists' lack of enthusiasm for pilot wave theory is related to the following methodological qualm: pilot wave theory, by proposing the guiding equation as a novel and distinctive law of nature, is a physical theory in its own right that should stand for itself. Most novel theories are proposed in response to *experimental* (and perhaps *inter-theoretical*) challenges, whereas pilot wave theory is designed to answer an essentially *foundational* challenge – the measurement problem – that arises in a quantum theory. Despite this difference, the methodological standards which apply to its evaluation are the same as in other evaluations of novel physical theories. And in these cases one normally hopes (or expects) that they are able to deliver solutions to *more than one* of the serious problems encountered by physics in the range of phenomena to which the theories in question apply. Pilot wave theory, however, does not really live up to such expectations. While it reproduces the predictive and explanatory successes of ordinary non-relativistic quantum mechanics, it does not help in solving any further empirical problems physicists are interested in solving. On the contrary, there are many predictive and explanatory successes of quantum theories pilot wave theory does *not* reproduce, since there are no fully fledged pilot-wave-theoretical counterparts to the quantum theories in question.

In particular, while there are some pilot-wave-theoretical versions of rather elementary bits of quantum field theory,[12] pilot wave theory currently remains far from being able to reproduce all the successful predictions of elementary particle physics, which are all essentially based on quantum field theory. In addition, there remains the serious challenge of reconciling the manifestly non-local character of pilot wave theory with the manifestly Lorentz invariant character of relativistic quantum field theory. Thus, it is presently unclear how pilot wave theory might be able to provide novel helpful perspectives and solutions to the problems not yet solved in fundamental physics such as, for example, the problem of combining the core insights of quantum field theory and gravity, the problem of accounting for the nature of dark matter and dark energy, the problem of accounting for the origin of the thermodynamic arrow of time, and the problem of accounting for the values of the constants of nature.

To expect from pilot wave theory that it may contribute to the solution of some of these long-standing and very difficult problems may appear to be a ridiculously tall order. However, if one is prepared to give some credit to the working physicist's perspective and consider the advantages and drawbacks of adopting pilot wave theory from that perspective, the order does not seem so unreasonably tall any more. All this is of course not meant to discourage future research into pilot-wave-theoretical versions of quantum field theories; quite the contrary. No doubt pilot wave theory is one of the most serious contenders among interpretations of quantum theory.

3.2.2 Alternatives: modal interpretations

Like pilot wave theory, *modal interpretations* belong to the class of approaches which solve the measurement problem by dropping assumption 1.A while keeping 1.B and 1.C. Just as pilot wave theory does, they dispense with collapse as an objective element of quantum theoretical time-evolution and treat measurements as ordinary interaction processes just like all others. The general idea behind modal interpretations is to replace the eigenstate/eigenvalue link by a different rule of which observables have determinate values under which conditions. Modal interpretations differ from pilot wave theory in that they generally do not attribute any privileged role to the position observable. Typically, in a modal interpretation, which observables have determinate values depends on the quantum state. Different modal interpretations vary as to *how* the quantum state determines the observables with determinate values.

The essential ideas behind modal interpretations go back to van Fraassen (see (van Fraassen [1974])) and references therein). Interest in them was spawned by the works of Kochen, Dieks, and Healey,[13] who independently arrived at similar strategies for singling out the observables with determinate values. Considering two-component composite quantum systems, all these attempts appeal to the *biorthogonal decomposition theorem*, which states that any vector $|\psi\rangle$ in the two-system tensor product Hilbert space can be written in the form

$$|\psi\rangle = \sum_i c_i |s_i\rangle \otimes |a_i\rangle, \tag{3.8}$$

with appropriate complex coefficients c_i and pairwise orthogonal vectors $|s_i\rangle$ and $|a_i\rangle$ from the systems' Hilbert spaces. The decomposition Eq. (3.8) is unique if the $|c_i|^2$ are all different. This observation may remind one of the state of the measured system together with the apparatus at the end of an ideal measurement process as displayed in Eq. (3.2), where one *would like* to obtain the result that the observables with determinate values are those with spectral decompositions in terms of $\{|s_i\rangle\}$ and $\{|a_i\rangle\}$, respectively. Motivated by this observation, modal interpretations which appeal to the biorthogonal decomposition propose that, if the state of a two-component system is $|\psi\rangle$ as decomposed in Eq. (3.8), the observables with determinate values are those with eigenvectors taken from the bases $\{|s_i\rangle\}$ and $\{|a_i\rangle\}$. This rule has the welcome consequence that in measurement contexts both the supposedly measured observable and the 'pointer' observable of the apparatus have determinate values.

For more complex combined systems and less idealised situations the appeal to the biorthogonal decomposition theorem alone does not suffice to single out determinate values as required by empirical adequacy. Further serious problems arise for less-than-ideal measurement interactions, where modal interpretations run the risk of not assigning determinate values to the observables for which they are required on empirical grounds. Finally, modal interpretations are confronted with severe difficulties when it comes to quantum field theory.[14] All in all, though the field of modal interpretations has remained a lively area of research, it seems fair to say that it has not yet produced a completely satisfying foundational account of quantum theory.

3.3 Schrödinger time-evolution not universal?

As explained in Section 2.2, the textbook account of quantum measurement sidesteps the difficulty of accounting for determinate measurement outcomes by invoking measurement collapse. According to the textbook account, collapse supposedly occurs whenever the system is measured, which means that it must reject Maudlin's assumption 1.B. Maudlin, more contentiously, ascribes this view to Bohr and the other proponents of the Copenhagen interpretation:

> The 'traditional' interpretation of quantum theory, if it is to be coherent, must be a non-linear theory. Bohr and the other founders explicitly rejected Einstein, Podolsky and Rosen's argument that quantum theory is incomplete (thus asserting 1.A), and they also insisted that macroscopic measuring devices be described by the language of classical physics (thus asserting 1.C, since in classical terms the pointer must point exactly one direction). So they must deny 1.B. It is no accident that von Neumann's (1955) classic presentation of the theory explicitly postulates collapses. (Maudlin [2011], p. 9)

Whether Maudlin is right in classifying 'Bohr and the other founders' as proponents of this solution (rejecting assumption 1.B) seems doubtful,[15] but, if one thinks that indeed *any* interpretation must declare either 1.A, 1.B, or 1.C to be false, it is clear that the Copenhagen interpretation is among those that deny 1.B.

The two main problems of the standard account of measurement, according to which collapse occurs whenever the system is measured, were already mentioned in Section 2.2: first, that collapse is in conflict with relativity due to its instantaneous, and hence non-local, character, and, second, that the vagueness and anthropocentric character of 'measurement' render this notion unsuitable to define under which conditions collapses occur.

While the first problem may be ignored if one is prepared (grudgingly perhaps) to question the scope and validity of relativity theory, the second problem calls for a more immediate response. Indeed, there are various different proposals on the market as to how collapse can be introduced (or a modified version of collapse that has all desired consequences), while dispensing with anthropocentric notions altogether. In what follows I review those proposals that seem to be most influential and briefly assess their respective problems and virtues.

3.3.1 From Wigner to Penrose

As we have seen, the main reason why the standard textbook account of measurement in terms of collapse is unsatisfactory is that 'measurement' is an anthropocentric, and hence non-fundamental, notion. However, while it seems difficult to doubt this for 'measurement', there are some philosophical outlooks according to which *mental* notions such as 'consciousness' may well qualify as referring to *fundamental* features of the world. These are dualistic accounts of mind and matter, and, based on them, one may hope that the role of 'measurement' in the textbook account can be taken over by some mental notion without running into the charge of anthropocentrism.

And indeed, in an influential analysis of measurement in quantum theory due to London and Bauer from 1939 the authors suggest that 'the consciousness of an "I"' may be able to effect the required collapse and to render the measurement outcomes well-defined:

> [I]t is not a mysterious interaction between the apparatus and the object that produces a new ψ for the system during the measurement. It is only the consciousness of an 'I' who can separate himself from the former function $\Psi(x,y,z)$ and, by virtue of his observation, *set up a new objectivity* in attributing to the object henceforward a new function $\psi(x) = u_k(x)$. (London and Bauer [1939], p. 252)

Along similar lines, Wigner argues many years later 'that the being with a consciousness must have a different role in quantum mechanics than the inanimate measuring device' and concludes that the Schrödinger equation cannot be universally valid and 'the quantum mechanical equations of motion cannot be linear' (Wigner [1967], p. 180), thus denying assumption 1.B. More recently, the existence of a link between consciousness and quantum state reduction along broadly Wignerian lines has been claimed by Henry P. Stapp (Stapp [2003]). However, given the heavily dualistic commitments of approaches such as those by Wigner and Stapp and the manifestly ad hoc character of their accounts of how consciousness affects the measurement process, it is not surprising that this remains a minority view.

While the approaches just mentioned locate the 'triggering mechanism' of collapse in the mind, others locate it in yet-to-be-discovered physical mechanisms that are going to be revealed only in future physical theories. Roger Penrose, for example, proposes an alternative perspective on collapse, namely one according to which its correct dynamics will be part of a future theory of quantum gravity.[16] Even though this is certainly an interesting suggestion, it remains highly speculative and, so far, it has not been worked out in any detail. In any case, it would seem rather strange if we had to wait for a successful theory of quantum gravity to reveal the correct account of collapse, while quantum theory, as we presently have it, works so well when applied to all those phenomena to which physicists are in fact applying it.

3.3.2   The GRW model

Among all accounts which deny assumption 1.B, the GRW model is probably the formally best worked-out one. In the GRW model, collapses take the form of random spontaneous localisation processes with respect to which it is precisely specified *which form* they have, as well as *where* and *how often* they occur. Localisation processes ('collapses') are assumed to occur randomly with uniform probability around 10^{-16} per particle per second. The suggested spatial 'form' of collapse is multiplication of the wave function by a Gaussian ($\sim \exp(-x^2/(2a^2))$) with width $a \approx 10^{-7}$ m, where these values are chosen 'by hand' in order to make the account compatible with the manifest fact that macro-objects have well-defined positions. The probability density for collapse at a space-time point q is chosen identical to the Born Rule probability density $|\psi(q)|^2$. This allows one to recover the original quantum theoretical predictions in the presence of spontaneous localisation processes.

A crucial element of the GRW model, which allows it to be compatible with the manifest empirical facts, is that for multi-component systems the collapse probability per second is dramatically increased as a consequence of the entanglement relation between the states of the individual component systems. Thus, while the wave functions of individual (approximately isolated) quantum systems almost always follow the Schrödinger equation and are typically widely spread out in space, those of macroscopic systems that consist of, say, $\sim 10^{23}$ particles are always tightly localised.

There are (at least) two fundamentally different ways to link the formalism of the GRW model to physical reality: the first identifies the spontaneous localisation processes with the objective events which together make up the history of the universe. The resulting account is known as the 'flash' version of GRW theory due to the 'flash-like' character of the events that it postulates. In this picture, according to Bell, who proposed it first, 'a piece of matter then is a galaxy of such events' (Bell [2004], p. 205). The second way of linking the GRW formalism to physical reality reads the wave function as an expression of the matter density, which evolves in time as described by the GRW law of time-evolution for wave functions.[17] Both versions have their numerous merits and drawbacks, which to discuss would be far beyond the scope of this work.[18]

The main problems of GRW theory are, as conceded by J. S. Bell (one of the approach's most enthusiastic supporters), 'the arbitrariness of the jump function, and the elusiveness of the new physical constants' (Bell [2004], pp. 208f.). It seems that, in a charitable interpretation, the GRW model is best conceived of as a *toy model* that sketches the lines along which a conceptually more satisfying solution of the measurement problem that denies assumption 1.B might be formulated. Just as for pilot wave theory, there remains the challenge of formulating a relativistic version of GRW theory. Even though some remarkable progress has been made in this direction in recent years for the 'flash' version of the theory, especially by (Tumulka [2006]), this challenge remains essentially unaddressed.

3.4   No determinate outcomes

Among the three assumptions 1.A, 1.B, and 1.C from which the measurement problem arises, assumption 1.C is the only statement that seems to express part of our everyday knowledge. Denying it is thus not an immediate option and means making a radical (some would say 'desperate') move. The only conceivable way of making this move is by simultaneously giving an account of why it *seems* to us as if measurements had determinate outcomes even though they do not.

The essential idea behind those accounts which adopt this route is that both the measurement outcomes and the quantum states that are assigned in practice are in some sense *relative* to the situations (or perspectives) of the agents who register the outcomes and assign the quantum states. Typically, these accounts acknowledge the existence of a *complete* wave function of the universe as a whole, which never collapses.[19] This idea was put forward as a serious proposal for the first time in the form of Everett's 'relative state formulation' (Everett [1935]). Acknowledging Everett's influence and importance, it is nowadays common to refer to all (or at least most) approaches which try to solve the measurement problem by rejecting assumption 1.C as versions of the *Everett interpretation*.

3.4.1 Defining branches

Versions of the Everett interpretation concur that there are different *branches* of reality with respect to which determinate measurement outcomes can be defined. Different versions of the Everett interpretations offer different accounts of the branches: some conceive of them as distinct *universes* (or 'worlds', hence the label '*many-worlds* interpretation'); others identify them with different *minds*, many of them associated with one and the same organism. Others again remain silent about the nature of the branches, arguing that the branches are best characterised indirectly as whatever they turn out to be in the light of Everett-interpreted quantum theory itself.

The most straightforward motivation for Everett-style interpretations is that they can accept the formalism of quantum theory as it stands, without adding any surplus theoretical elements such as equations of motion for particle configurations or dynamics of collapse. David Wallace, in particular, repeatedly emphasises this point:

> If I were to pick one theme as central to the tangled development of the Everett interpretation of quantum mechanics, it would probably be: *the formalism is to be left alone*. What distinguished Everett's original paper both from the Dirac-von Neumann collapse-of-the-wavefunction orthodoxy and from contemporary rivals such as the de Broglie-Bohm theory was its insistence that unitary quantum

mechanics need not be supplemented in any way (whether by hidden variables, by new dynamical processes, or whatever). (Wallace [2007], p. 311)

But how does introducing the concept of branches of reality help reconcile a formalism according to which measurement outcomes (and other macroscopic events) are objectively indeterminate with the fact that measurement outcomes appear always determinate?

The first challenge which any version of the Everett interpretation has to answer is to individuate the desired branches and to explain how they arise. The idea which underlies most contemporary responses to this challenge is to rely on environment-induced decoherence as the mechanism which effectively generates the desired branching structure. As explained above in Section 3.1, it selects (at least approximately) a preferred basis in the sense that the resulting reduced density matrix

$$\rho_{\text{post},S} = \sum_i |c_i|^2 |s_i\rangle\langle s_i| \qquad (3.9)$$

for the system (and analogously the density matrix of the system and apparatus taken together; see Eq. (3.6)) is a mixture of states $|s_i\rangle\langle s_i|$ which are elements of a basis selected by the interaction between the system and its environment (which can be the measuring apparatus). In other words, the quantum state of the measured system does not contain any superpositions of different states from the environment-selected basis. Most modern Everettians regard this observation as the clue to the insight that environment-induced decoherence is to be regarded as the mechanism that creates and thus defines the branches: they are simply identified with those elements of the environment-selected decoherence basis which appear with nonzero coefficients c_i in Eq. (3.9).

To regard environment-induced decoherence as inducing the branch structure leads to various further questions. For example, an important feature of the branches, so defined, is that they have only fuzzily defined criteria of identity as a consequence of the fact that environment-induced decoherence is always *incomplete* in practice: it is only to some (good but limited) approximation that Eq. (3.9) correctly describes the reduced density matrix for a system subjected to real-life decoherence. As a consequence, the resulting branches do not have any sharply defined identities and the question of how *many* such branches there are has no clear-cut answer. Critics of the Everett interpretation, notably Kent (see his contribution to (Saunders et al. [2010])), regard fuzzily defined branches as incapable of bearing the weight that is placed on them in the Everett interpretation. Defenders of the Everett interpretation respond by arguing that the branches need only have approximately defined identities to fulfil the role they are supposed to fulfil.

3.4.2 How do probabilities fit into the picture?

By most accounts, the most serious problem for the Everett interpretation is to reconcile the overtly *indeterministic* and, as it seems, irreducibly *probabilistic* character of quantum theory in its applications with the fact that the time-evolution of the Everettian universal wave function evolves in a completely deterministic way, such that all possibilities are equally realised and probabilities do not seem to enter the picture at all. Maudlin emphasises this difficulty for the Everett interpretation in his exposition of the measurement problem discussed above when he formulates a second version of the third assumption used to derive the measurement problem that reads as follows:

> 2.C Measurement situations which are described by identical initial wave-functions sometimes have different outcomes, and the probability of each possible outcome is given (at least approximately) by Born's rule. (Maudlin [1995], p. 11)

Everettians thus need to give an account of why not only assumption 1.C, but also assumption 2.C, seems to hold. It is here where the Everettian interpretation appears to run into the most serious trouble.

To see the problem in its most striking form, we may look at the Everettian definition of the branches as the elements of the environment-selected decoherence basis which have nonzero coefficients c_i in Eq. (3.9). According to this individuation condition, it is completely irrelevant *which* (nonzero) value some coefficient c_i actually has. However, in the actual *application* of quantum theory, the numerical values of the coefficients c_i are highly relevant, for the $|c_i|^2$ are interpreted as probabilities.[20] How can Everettians respond to this challenge?

The idea to which they typically resort in response to it is to conceive of the quantum probabilities as expressing the agents' subjective ignorance as regards the *self-locating* information concerning which branch they inhabit (or will inhabit). There are different accounts as to how the specific numbers $|c_i|^2$ may come into play here as expressing their appropriate self-locating beliefs. These accounts range from those which simply *postulate* that a subject's rational credence with respect to the possibility of finding herself in the branch associated with $|s_i\rangle$ is $|c_i|^2$ (this approach is proposed by Lev Vaidman (Vaidman [1998])) to others which attempt to *derive* the Born Rule from the quantum theoretical formalism using elements of decision theory (see (Deutsch [1999]) and (Wallace [2007])). Both proposals, the one which simply postulates that the $|c_i|^2$ are the relevant probabilities and the one which aims to derive this from something more fundamental, are highly controversial. The main criticism of the first approach is that it is not an option to *postulate* what the agents' rational credences are, unless there is something else in the branching structure of the Everett universe that suggests it. The main criticism of the second approach is that our preferences might be different than assumed in the decision-theoretic arguments without in any way being irrational.[21]

To conclude, the problem of defining branch structure and in particular that of giving an account of probabilities in a branching universe provide serious challenges for the Everett interpretation. There are no universally accepted solutions to these problems.

In the next chapter I turn to a very different approach of 'leaving the formalism alone',[22] namely one which is based on a *therapeutic* methodology for approaching philosophical problems.

Part II

The Rule Perspective

4
Motivating a Therapeutic Approach

4.1 Absence of the foundational problems in practice

As explained in the previous section, the currently most-discussed interpretations of quantum theory can be classified according to how they (try to) solve the measurement problem. Accounts which are based on what I call the 'therapeutic' type of approach to philosophical problems propose a radically different strategy to overcome the measurement problem. The present chapter presents the essential ideas behind therapeutic approaches, briefly sketches their background in the history of philosophy, and explains how one might try to put them to work in the foundations of quantum theory.

There is at least one prima facie reason for being sceptical as to whether the measurement problem (and, similarly, the problem of quantum non-locality) is indeed the profound conceptual difficulty the suggested interpretations take it to be, namely, that it is suspiciously absent from quantum mechanical practice. In particular, the measurement problem does not become manifest as a serious limitation of physicists' predictive and explanatory resources, even when it comes to experimental contexts where 'measurements' are abundant or when it is not really clear whether some process deserves to be regarded as a 'measurement interaction' in whatever precise sense of the term. In fact, the measurement problem's lack of impact on quantum theoretical practice is what makes it a *foundational* problem par excellence. Even John Bell, one of the staunchest critics of textbook quantum theory, acknowledges that the theory is fine *for all practical purposes* ('FAPP') (Bell [2004], p. 33). But while this makes the foundations of the theory no less unsatisfying for Bell, an alternative reaction is to regard the absence of the measurement problem in practice as an indication that it is not a genuine problem at all. Perhaps, one might suspect, it is merely a *pseudoproblem* that arises from a distorted perspective on how the quantum theoretical formalism connects to physical reality.

As announced in the introductory chapter, the present work takes this idea seriously and explores the possibility that the foundational problems may vanish after certain mistaken conceptual presuppositions as regards the nature of quantum states have been corrected. The idea, in other words, is to *dissolve* (rather than solve) the paradoxes by proposing a perspective on how quantum theoretical language relates to the world according to which at least one of the assumptions necessary to derive the paradoxes is not so much wrong as rather conceptually ill-formed. The motivation underlying this approach may be called 'therapeutic' inasmuch as it aims at 'curing' us from what it sees as unfounded worries based on conceptual misunderstandings. The following section gives a brief account of the origins and scope of the conception of philosophical investigation as 'therapeutic' that informs this approach; the next-to-following section gives an outline of why adopting a version of the *epistemic* conception of quantum states seems to be the most promising step for putting the suggested 'therapeutic' ambitions to work in quantum theory.

4.2   Philosophy as therapy

The idea that philosophical reflection may play a therapeutic role has a long tradition. It is part of the teachings of the Hellenistic philosopher Epicurus and his follower Lucretius, both of whom conceive of philosophical reflection as a therapeutic means to free ourselves from our fear of death and the gods. However, the approach to the foundations of quantum theory taken here is 'therapeutic' in a different sense: the problems and puzzles on which it focuses and which it tries to dissolve are primarily intellectual and do not concern the individual and its emotional balance as a whole.

One important figure in the history of modern philosophy is associated especially closely with the conception of philosophy as therapy that is at stake in the present work. This is the later Wittgenstein, who, comparing philosophical methods to 'different therapies' (Wittgenstein [1958] § 133), writes that '[t]he philosopher's treatment of a question is like the treatment of an illness' (Wittgenstein [1958] § 255). Wittgenstein critically derides philosophers' attempts to formulate philosophical *theories* which supposedly solve and address philosophical problems in a similar vein to how scientific theories solve scientific problems. For him, both the aims and the methods of philosophy are essentially different from the aims and methods of the sciences. Contrary to the sciences, philosophy neither *discovers* nor *explains* anything that is not in principle open to view.[23] Rather, philosophy should strive to reveal and remove the conceptual misunderstandings which create the (misleading) impression that there are philosophical problems in analogy to scientific ones, which would require philosophical theorising. As Wittgenstein formulates it in the *Philosophical Investigations*:

> [T]he clarity that we are aiming at is indeed *complete* clarity. But this simply means that the philosophical problems should *completely* disappear. The real discovery is the one that makes me capable of stopping doing philosophy when I want to. The one that gives philosophy peace, so that it is no longer tormented by questions which bring itself in question. (Wittgenstein [1958] § 133)

The proper means to achieve 'complete clarity', according to Wittgenstein himself, is to pay careful attention to our actual linguistic practices in which the notions giving rise to the supposed philosophical problems are at home and successfully used. This type of clarity is evidently very different from the clarity (in the sense of 'precision') that can only be found in a mathematical formalism. For Wittgenstein, only the clarity that is achieved by focusing on our actual practices and employment of language is the genuinely philosophical one.

In the opening paragraphs of (Wittgenstein [1958]), Wittgenstein provides evidence of the richness and vastness of linguistic phenomena, which he sees as strongly suggesting that no uniform philosophical account of such notions as 'language' and 'meaning' is to be had. One particularly widespread type of 'disease' that he diagnoses is the assumption, made by many philosophers, that the descriptive, 'fact-stating' mode of language is the systematically fundamental one and that assertion and description are the paradigm modes of employment of language. According to Wittgenstein, in contrast, the descriptive use of language is merely one among many and should not be regarded as in any way privileged.

A mistake made by many philosophers in different contexts, according to Wittgenstein, is to assume, without further question, that some body of language is descriptive and then jump to conclusions about what that body of language describes. Wittgenstein traces various philosophical problems to this mistake and proposes philosophical therapy in the sense just discussed as the appropriate remedy. This diagnosis plays a role in various cases of philosophical interest to which Wittgenstein applies his therapeutic ambitions, for instance in his analysis of discourse involving mental notions (including the famous 'private language argument' in ([Wittgenstein, 1958, §§ 243–258])) and in his remarks on the language of mathematics. Highlighting the expressive dimension of language involving mental notions (while not falling into naive expressivism) he suggests that mental vocabulary is neither to be construed as describing private 'internal' processes nor – as the behaviourist would have it – public 'external' ones. In the case of mathematics, Wittgenstein strongly objects to the widespread view of mathematical language as describing an abstract and eternal 'mathematical reality' and proposes, on the positive side, a radically different perspective on mathematical sentences as conceptual *norms* which govern our usage of mathematical concepts and their applications.[24]

Philosophers inspired by Wittgenstein's therapeutic approach to philosophical problems have adopted and applied it to topics beyond Wittgenstein's own philosophical interests. Paul Horwich, for instance, proposes an approach to epistemological problems which he dubs 'therapeutic Bayesianism', 'whose primary goal is the solution of various puzzles and paradoxes that come from reflecting on scientific method' (Horwich [1993], p. 607). In a similar spirit are the various *deflationary* approaches to philosophical problems, which are nowadays championed by various different philosophers in different areas of philosophy. As examples, one may mention deflationism about 'truth' (defended, for instance, by Paul Horwich, Hartry Field, and Huw Price; see (Stoljar and Damnjanovic [2012]) for a useful overview), William Tait's deflationism concerning mathematical truth and existence (Tait [2001]), and Arthur Fine's deflationary stance in debates about realism and antirealism in philosophy of science, which he himself calls 'NOA', the 'Natural Ontological Attitude' (Fine [1996]).

In a spirit akin to that which underlies these deflationary approaches in other parts of philosophy, the present work probes the working hypothesis that the paradoxes of quantum theory may not require a modification of the theory or a radical re-interpretation (in the sense of the many-worlds interpretation, say). Rather, in line with the therapeutic conception of philosophy, the idea to be pursued is that the paradoxes arise from conceptual misunderstandings which one may hope to 'cure' by philosophical therapy. In the following section, I present some more specific thoughts as to what those misunderstandings might be.

4.3 Quantum states as non-descriptive

As explained in the previous section, the kind of misunderstanding which, according to Wittgenstein, creates many (or even most) philosophical (pseudo-) problems is that one tends to presuppose – whether implicitly or explicitly – that the elements of language in which these problems are formulated is descriptive, while their actual use is much more complicated. And indeed, with respect to the formalism of quantum theory – in particular as regards quantum states – the assumption that their mode of use is descriptive (in the sense of *representing* features of reality) is tacitly made in almost all philosophical treatments.

For example, in the passage quoted in Section 3.4.1 from (Wallace [2007]), where Wallace emphasises that the core idea of the Everett interpretation is that the formalism is to be 'left alone' in that nothing needs to be added to it, the very idea of 'leaving the formalism alone' is much less straightforward than Wallace suggests. Taken by itself, the formalism is just an uninterpreted piece of mathematics. Its extra-mathematical significance exists only in virtue of its application by competent physicists to specific problems of physical interest. Wallace's case for the Everett interpretation as the single outstanding take on quantum theory which does not 'supplement' the formalism 'in any way' is valid only if one presupposes that the role of quantum states is to represent the physical facts. And indeed, making this assumption is a natural first step when interpreting quantum theory, and to outline its implications is both highly important and rewarding, as the work of Everettians impressively demonstrates. However, for those impressed by the problems encountered by Everettians when trying to 'leave the formalism alone' while giving it a descriptive reading, a natural next step is to look for alternatives which also 'leave the formalism alone', while not construing quantum states as describing reality.

In accordance with the received terminology found in the literature, I refer to accounts which deny that the proper linguistic function of quantum states is to represent features of reality as *non-ontic* accounts of quantum states. These contrast with *ontic* accounts, which assume that, for any quantum system, there exists precisely one *true* quantum state it 'is in'. Note that in order to qualify as 'ontic' in this sense, an account need not assume that the quantum state a quantum system is in describes it *completely*. Ontic accounts (in the sense used here) may well assume that the complete physical state λ of a physical system comprises more variables than its quantum state ψ (or ρ), in which case λ can be written as $\lambda = (\psi, \omega)$. Furthermore, ontic accounts of quantum states need not ascribe *metaphysical reality* in any strong sense to the quantum states. An account qualifies as 'ontic' in the sense used here if it accepts a literal reading of the notion of a quantum state a quantum system 'is in'.

Coming back to the idea of approaching the foundations of quantum theory in a therapeutic spirit, the most natural strategy seems to explore what it takes to adopt a non-ontic account of quantum states. What are the prospects of avoiding the measurement problem by making this move?

4.3.1 Dissolving the measurement problem

Let us assume that the linguistic function of quantum states is not that of representing features of physical reality in the sense that the notion of a quantum state a quantum system 'is in' is misleading. Now, whatever exactly may come to replace it, it is evident that by denying the idea that systems 'have' quantum states or 'are in them' one denies an important presupposition of Maudlin's assumption 1.A, namely that a quantum state specifies any 'physical properties of a system' in the first place. This presupposition is a shared assumption among all the solutions to the measurement problem discussed in the previous chapter, both those which accept assumption 1.A and those which reject it.

Adopting a non-ontic account of quantum states and denying that quantum states represent features of reality undermines the conceptual presuppositions of assumption 1.B as well, simply because this involves rejecting the very notion of '*the* wave function' of a system, not to speak of that wave function's evolution in time. Agents who are competent in applying quantum theory do of course *assign* quantum states to quantum systems and they *make them* undergo unitary time-evolution as well as collapse, but acknowledging this is very different from interpreting unitary time-evolution and collapse as time-dependent physical features of quantum systems evolving through time.

With respect to assumption 1.C, in contrast, adopting a non-ontic account of quantum states has no ramifications. The trivial fact that measurements have determinate outcomes is not put into doubt.

Even though non-ontic accounts of quantum states do not give rise to the measurement problem in the formulation suggested by Maudlin, they do not give any guarantee that other problems might not appear in its place. One objection critics are likely to raise, for instance, is that rejecting the conceptual presuppositions of the measurement problem does not help us account for why measurements have determinate outcomes and makes it no less mysterious. No-go theorems on assigning determinate values to observables (the Kochen–Specker theorem, in particular; see Appendix B) seem to make it unattractive to assume that all observables have determinate values at all times, so the question arises as to why those observables which we measure never fail to exhibit determinate values when we measure them. And indeed, it is true that by dissolving the measurement problem in the sense of a sharp antinomy of conflicting assumptions, I have not been giving an account of why those processes we refer to as 'measurement processes' yield determinate outcomes in the first place. However, different ways of spelling out the idea that quantum states have a non-descriptive function respond differently to this latter challenge. A response to the challenge of specifying why measured observables never fail to have determinate values from the point of view of the non-ontic account of quantum states to be developed in Chapters 5 and 6 will be given in Chapter 9.

In what follows, I argue that the idea that quantum states have a non-representational linguistic function can be fleshed out most promisingly by regarding them as in some sense (to be made more precise later on) reflecting the epistemic conditions of those who assign them. From such a perspective, arguably, measurement collapse starts to make sense.

4.3.2 Collapse as update

Let us assume, in conformity with the considerations presented in the previous section, that quantum states do not represent features of physical reality, but reflect, in some way to be specified, the epistemic relations of the agents who assign them to the systems they are assigned to. This *epistemic* conception of quantum states immediately suggests a straightforward reading of collapse as reflecting a change in the agent's epistemic relations to the system as a consequence of having registered the measured result. And indeed, the idea that collapse corresponds to a shift in what the agent knows of the system occurs to almost any student of quantum mechanics when she learns the collapse postulate. Although this reading of collapse is not normally considered to be part of the Copenhagen interpretation, it has been endorsed by some of the thinkers most closely associated with that view.[25] Perhaps the most important example of such tendencies can be found in Heisenberg's writings, where collapse is characterised as representing a discontinuous change in the knowledge of the agent:

> Since through the observation our knowledge of the system has changed discontinuously, its mathematical representation also has undergone the discontinuous change and we speak of a 'quantum jump'. (Heisenberg [1958], p. 28)

Indeed, if one thinks that quantum states are properly assigned only if they somehow *accurately* reflect the epistemic situations of those who assign them, one is not only able to *accommodate* collapse, but even to *justify* it. For its role and function is to bring the state assigned to the system in agreement with incoming information as regards what values the observables of the system currently have (or have had). For instance, if an agent assigning a quantum state learns that the value of an observable A being measured at t_0 lies, at that time, between the two values a_0 and a_1, the state assigned for times t immediately after t_0 seems to be *obligated* to ascribe probability 1 (or at least very close to 1) to the value of A lying between a_0 and a_1. If the pre-measurement state does not fulfil this requirement, it does not properly reflect what the agent knows of the system and must therefore be adjusted to what she knows by applying a projection to it. To conclude, the complaint that measurement collapse is, as Ruetsche puts it, 'a Humean miracle, a violation of the law of nature expressed by the Schrödinger equation' (Ruetsche [2002], p. 209) is compelling and natural only if one thinks of the quantum state and its time-evolution as representing what happens to the system. If, more in line with the original motivation of invoking collapse, one regards it as reflecting a sudden change in the state-assigning agent's epistemic situation, this changes completely and collapse starts to look very natural.

4.3.3 Types of epistemic accounts

One of the most outspoken proponents of the epistemic conception of quantum states – sometimes associated with the Copenhagen interpretation – is Rudolf Peierls, a former PhD student of Heisenberg and an eminent figure in 20th-century physics in his own right. According to him, the quantum state 'represents our *knowledge* of the system we are trying to describe', which implies that the quantum states assigned by different observers 'may differ as the nature and amount of knowledge may differ'.[26] This provides us with a very useful necessary condition for accounts to qualify as versions of the epistemic conception of quantum states: they must allow that different agents may legitimately assign different quantum states to the same system and, therefore, must be non-ontic. If there existed any such thing as an agent-independent 'true' quantum state of a quantum system, assigning precisely that state would be the one and only objectively correct way of assigning a quantum state to the system and assigning any other state would be wrong, independently of one's epistemic relation to it.

Epistemic accounts of quantum states come in two different types: those on which the focus of the present work lies are motivated by the therapeutic approach to philosophical problems, where the hope is that by adopting the epistemic conception of quantum states and spelling it out in a convincing way one can get around the paradoxes of measurement and non-locality without formulating any more fundamental *theory* about underlying ontic states. Conceptual clarification in the sense of the therapeutic approach alone, it is hoped, may lead to a perspective on quantum mechanics according to which this theory is fine as it stands and as actually applied by working physicists. According to these accounts, no further interpretive take on quantum mechanics, be it in terms of hidden variables, branching worlds, dynamics of collapse, or whatever else, is needed in addition to the epistemic conception of states.[27]

Epistemic accounts of the second type combine an epistemic account of quantum states with a theoretical account of additional (ontic) variable configurations which are added to the standard formalism of quantum theory. The defining characteristic of these theories[28] is that an ontic state is compatible with several quantum states ψ. Important contributions to this type of approach include the construction of an explicit toy model based on the epistemic conception of quantum states by (Spekkens [2007]), where Spekkens demonstrates that many of quantum mechanics' most typical and astonishing features can be reproduced in a simple toy theory together with a principle that restricts the degree of knowledge an agent may have of the ontic state of the system. More recently, a theorem due to Pusey, Barrett and Rudolph (Pusey et al. [2012]) has stimulated a lot of interest and research activity, which establishes that an important class of hidden-variable models based on the epistemic conception yields predictions that are necessarily incompatible with those of standard quantum theory. Since accounts of the second type are not based on any therapeutic ambitions, the strategical starting point of this discussion is with those of the first type. However, the distinction between the two types of epistemic accounts turns out not to be sharp in the end. Chapter 12 elaborates on this issue when discussing in what sense the idea of a complete description of physical reality is available on the specific epistemic account of quantum states that is going to be developed in the meantime.

4.4   Space-time structure and collapse

In this section I argue that epistemic accounts of quantum states not only dispel the measurement problem (in the formulation discussed in Chapter 3) but also remove some of the mystery concerning quantum non-locality and entanglement.

As discussed in Section 2.3, part of the worry of whether quantum theory is compatible with relativity theory rests on the challenge that the time-evolution of quantum states including collapse in whatever suitable form seems impossible to reconcile with the requirement of Lorentz invariance as inferred from relativity theory. Let us briefly recapitulate the simple example used to illustrate this challenge in Section 2.3: consider a two-component system in an EPR-Bohm setup, where two spin-1/2 particles 1 and 2 are prepared according to a preparation procedure that requires assigning an entangled state, for instance, the state $|\psi_{EPRB}\rangle = \frac{1}{\sqrt{2}} \left(|+\rangle_1 |-\rangle_2 - |-\rangle_1 |+\rangle_2 \right)$, for their combined spin degrees of freedom. Assume that the two systems have been brought far apart and an agent Alice, located at the first system, measures spin in a certain direction. Having registered the result and having applied the projection postulate, she assigns two no longer entangled states to the systems 1 and 2, which depend on the choice of observable measured and on the measured result. Another agent, Bob, located at the second system 2, may also perform a spin measurement (in the same or in a different direction of spin) and proceed to assign a pair of no longer entangled states, determined by the projection postulate, to the two systems in an analogous way.

From the point of view of the ontic conception of quantum states the challenge of reconciling quantum theory including collapse with relativity theory is formidable here due to the fact that collapse is instantaneous, whereas there is no absolute notion of simultaneity with respect to which instantaneous collapse could be defined in the relativistic setting. The most natural move may seem to be to single out a suitable preferred frame of reference with respect to which instantaneous collapse could occur at a single simultaneous instant after all. One might hope that such a preferred frame arises in a natural and *dynamical* way. However, as Maudlin argues with great care (Maudlin [2011], pp. 185–187), all the ideas which come to mind as to how the preferred frame might arise are beset with severe conceptual difficulties. Most of them attribute a special role to the *centre of mass* – either of the entangled systems themselves, their 'source' (if definable), or the remaining amount of matter in the rest of the universe –, and the problem with any such proposal is that the very notion of a centre of mass presupposes that of unambiguous simultaneity (since the centre of mass is computed in terms of the positions of masses *at a given time*).

In addition, and as mentioned in Section 2.3, the prospects for singling out a preferred reference frame are particularly dim with respect to situations where the two measurements carried out on distant systems prepared such that an entangled state must be assigned are each performed first in their own rest frame. (Zbinden et al. [2001]) In that case it seems especially hopeless to give a non-arbitrary answer to the question as to which measurement occurs first and triggers the abrupt change of state which afflicts the other.

The approach towards reconciling measurement collapse with the space-time structure of special relativity which – among those which preserve the ontic conception of quantum states – seems to be the least hopeless is to make the time-evolution of quantum states dependent on foliations of space-time into sets of parallel hyperplanes. In this setting, advocated, for instance, by Fleming and Myrvold,[29] collapses occur on (and are relative to) space-like hyperplanes (or, in curved space-time, space-like *hypersurfaces*), which means that they are taken to be relative with respect to the reference frame just as simultaneity relations between events in special relativity are. While formally consistent, this programme has been sharply attacked from a philosophical point of view by Maudlin, who argues that it entails an extreme form of holism by making *any* property dependent on hyperplane and thus rules out intrinsic properties for limited space-time regions, thereby breaking with the Einsteinian ideal of separability in a particularly radical fashion.[30] A further problem with taking probabilities to be hyperplane-dependent, to be presented in Chapter 10 of this work, is that the hyperplane-dependent 'probabilities' do not really qualify as genuine probabilities if one accepts David Lewis' famous *Principal Principle* (Lewis [1986a]), since they do not impose any constraints on rational credences for all agents situated along space-like hyperplanes (See Section 10.3 for details).

The epistemic conception of quantum states undermines the conceptual presuppositions of the reasoning that leads to this problem by rejecting the notion of a quantum state a quantum system is in and by interpreting the assignment of different states to the two systems by the different agents as at once legitimate and very natural: Alice knows about the preparation procedure for the combined two-particle system, and when she registers the result pertaining to her own system this affects her epistemic condition with respect to the second. The state she assigns to it reflects her epistemic relation to it, and there is no need to assume that her measurement of her own system physically influences the second. The same considerations apply for Bob. Predictions for the results of measurements derived on the basis of entangled states may still be baffling and unexpected, but the dynamics of quantum states do not give rise to any incompatibility between the principles of relativity theory and those of quantum theory as construed by the epistemic conception of states. The paradox of quantum non-locality (inasmuch as it resides in the challenge of reconciling the time-evolution of quantum states with the principles of relativity theory) is *dissolved* here by rejecting two of its core conceptual presuppositions, namely, the ontic conception of states and the associated interpretation of quantum theoretical time-evolution as a physical process.

As in the case of the measurement problem, the dissolution of the paradox in its original and most widely cited form gives no guarantee that the problem may not appear elsewhere in a different form. And indeed, with respect to quantum non-locality the true challenge is often believed to arise from the fact that quantum theory violates Bell's criterion of *local causality*: a criterion which is meant to implement the idea – motivated by considerations on relativistic space-time – that causal influences do not travel faster than light and the probabilities of events depend only on what occurs in their backward light cone. However, as I shall argue in Chapter 10, this challenge fails, because local causality, properly construed, is not violated in quantum theory at all. In the next chapter, the discussion turns to how the epistemic conception of quantum states should be spelled out in detail.

5
In Search of a Viable Epistemic Account

5.1 Knowledge of probabilities versus probabilities as degrees of belief

A quantum state, via the Born Rule, assigns probabilities to the different possible values of observables of a quantum system. It is only in virtue of these probabilities – or, alternatively, expectation values – that quantum mechanics is empirically testable. Since assigning a state to a system means assigning probabilities to the values of observables, it may seem natural to read Peierls' claim that the state 'represents our knowledge of the system' (Peierls [1991], p. 19) as a shorthand for saying that the state represents our knowledge of these probabilities. The latter view is sometimes even straightforwardly identified with the epistemic conception of states, for example by Marchildon, who claims that '[i]n the epistemic view [of quantum states], the state vector (or wave function or density matrix) does not represent the objective state of a microscopic system [...], but rather our knowledge of the probabilities of outcomes of future measurements' (Marchildon [2004], p. 1454). However, according to an argument due to Fuchs and endorsed by Timpson,[31] the idea that quantum states represent our knowledge about quantum probabilities should not be regarded as the core claim of the epistemic conception of quantum states, simply because, as they argue, this idea is in fact *incompatible* with the latter.

The reason for this is the so-called factivity of knowledge, namely that, according to what 'knowledge' means, it is impossible to know that q unless 'q' is indeed true. Due to this elementary fact about 'knowledge', the argument goes, the view described by Marchildon – that the state reflects our knowledge of probabilities – seems to be incompatible with what we determined to be an essential ingredient of the epistemic conception of states, namely the assumption that different agents may legitimately assign different states to one and the same system. For assume that probabilities are indeed the objects of our knowledge so that an agent might know the probability p of a certain measurement outcome E to occur. In this case an immediate consequence of the fact that 'knowledge' is factive seems to be that p is the one and only correct, the *true*, probability of E occurring. Since this line of thought applies for any possible measurement outcome E some probability p is ascribed to by the state assigned to the system, we seem bound to conclude that the probabilities obtained from this state are the (only) true ones so that any other assignment of probabilities would simply be wrong. But this conclusion would be incompatible with the fundamental assumption of the epistemic conception of states that the states assigned by different observers, as Peierls writes, 'may differ as the nature and amount of knowledge may differ'.

To evade this argument, the proponent of the epistemic conception of quantum states who defends the idea that quantum states are expressions of our knowledge about quantum probabilities may offer an agent-relative account of quantum probabilities and argue that the probabilities to be *known* by the different agents may differ. This move does not really avoid the sketched difficulty, however, as the challenge remains to account for what the actual (agent-relative) probabilities are which we may or may not know. Unless more is said as to how these probabilities supposedly depend on the epistemic situations of the assigning agents, what we have is an *ontic* account of quantum probabilities and, thereby, an ontic account of quantum states. Therefore, Fuchs and Timpson seem to be essentially right when they reject the idea that quantum states represent our 'knowledge of' quantum probabilities as not very helpful for spelling out the epistemic conception of states.

5.2 Quantum Bayesianism

5.2.1 Probabilities as subjective

In response to the difficulties involved in conceiving of quantum states as representing knowledge of probabilities, a version of the epistemic conception of quantum states has been proposed which conceives of quantum probabilities as *subjective* degrees of belief in the sense of the subjective Bayesian interpretation of probability. This account, known as *quantum Bayesianism*, has been worked out in great detail by Fuchs,

Caves, and Schack. (Fuchs refers to the position as worked out in the light of his personal philosophical preferences as 'QBism'; see (Fuchs [2010]).) The essence of quantum Bayesianism is to regard quantum probabilities as reflecting not our knowledge, but instead our *degrees of belief* as regards possible outcomes of what are traditionally called 'measurements', referred to by Fuchs as 'our interventions into nature' (Fuchs [2002], p. 7).

Degrees of belief may legitimately differ from agent to agent without any of them being irrational or making any kind of mistake. So, from the quantum Bayesian point of view different agents may indeed legitimately assign different quantum states to the same system, the notion of a quantum state as a state some quantum system 'is in' is rejected, and the position qualifies as a potentially consistent epistemic account of quantum states. Fuchs expresses this in his dictum that 'quantum states do not exist':[32] they are not part of the furniture of our world.

Quantum Bayesianism, as conceived by its inventors, is not merely a novel philosophical perspective on quantum theory but also an ambitious programme of reformulating quantum theory in terms of probabilistic and information-theoretic notions rather than abstract mathematical ones. Such an information-theoretic reformulation, it is hoped, might help us understand which elements of quantum theory represent physical features of the world and which others are mere reflections of what constitutes rational reasoning of the agents using the theory. Fuchs announces the following strategy for his programme of determining the objective core of the world according to quantum theory – comparable, according to him, to the space-time manifold in general relativity as described in a subject-independent and completely coordinate-free way: 'Weed out all the terms that have to do with gambling commitments, information, knowledge, and belief, and what is left behind will play the role of Einstein's [space-time] manifold', namely, as he explains, 'a mathematical object, the study of which one can hope will tell us something about nature itself, not merely about the observer in nature' (Fuchs [2002], p. 6). In what follows I shall not be concerned with the prospects and problems of the quantum Bayesian reformulation programme of quantum theory, even though it certainly merits the attention it gets, but rather focus on the assumptions about the status of quantum observables, states, and probabilities on which it rests.

One of the greatest achievements of quantum Bayesianism is its demonstrated ability to make sense of talk about 'unknown quantum states', even though, from the quantum Bayesian point of view, the notion of an 'unknown quantum state' makes no sense. This notion features prominently, in particular, in a discipline called 'quantum state tomography', the official business of which is to determine the true unknown states quantum systems are in. Quantum Bayesians manage to make sense of what is done in quantum state tomography by invoking a variant of de Finetti's classic representation theorem on sequences of events that are subjectively judged to be 'exchangeable' (Caves et al. [2002]).

Based on this theorem, it becomes clear why different agents who register the same measured data will, step by step, come to agree in their state assignments – even if there is no true unknown state the system is actually in and even if the states they originally start out with are very different. The sole presupposition for this to happen is that the assigning agents (subjectively) judge the states of the sequence of measured systems to be *exchangeable*, roughly meaning that for them both the individual positions of the systems in the sequence of trials and the number of measurement trials carried out play no role.[33] As demonstrated by Caves, Fuchs, and Schack, this assumption is sufficient to enforce that 'the updated probability $P(\rho|D_K)$ becomes highly peaked on a particular state ρ_{D_K} dictated by the measurement results, regardless of the prior probability $P(\rho)$, as long as $P(\rho)$ is nonzero in a neighbourhood of ρ_{D_K}' (Caves et al. [2002], p. 4541). Hence, this theorem can be used to explain why the states assigned by different agents starting from different priors will become practically indistinguishable after a sufficiently large number of experiments witnessed without there being any such thing as the 'unknown state' any of the systems really is in.[34] As this shows, taking quantum theoretical practice seriously inasmuch as it involves talk about 'unknown quantum states' does not mean that one has to embrace an ontic account of quantum states in which such talk is interpreted literally.

5.2.2 Values of observables as subjective?

Quantum Bayesianism goes extremely far in characterising elements of the quantum mechanical formalism as subjective in order to be consistent as an epistemic account of states. How radical the view really is becomes strikingly clear from the fact that, for any given measurement device, quantum Bayesianism denies the existence of a determinate answer to the question of *which* observable is measured in that setup. As explained by Fuchs:

Take, as an example, a device that supposedly performs a standard von Neumann measurement $\{\Pi_d\}$, the measurement of which is accompanied by the standard collapse postulate. Then when a click d is found, the posterior quantum state will be $\rho_d = \Pi_d$ regardless of the initial state ρ. If this state-change rule is an objective feature of the device or its interaction with the system – i.e., it has nothing to do with the observer's subjective judgement – then the final state must be an objective feature of the quantum system. (Fuchs [2002], p. 39)

Fuchs' reasoning can be summed up as follows: if it is an objective feature of the device that it measures an observable having a set $\{\Pi_d\}$ of one-dimensional projection operators as its spectral decomposition, then, as soon as a 'click d' has been registered, the state that must be assigned after measurement is Π_d independently of the state ρ which has been assigned before. No freedom of state-assignment remains in this case and we seem to have ended up with an agent-independent true post-measurement state Π_d the existence of which is incompatible with the epistemic conception of states we wanted to spell out in detail. This motivates Fuchs' conclusion that in consistently spelled out versions of the epistemic conception of quantum states there can be no objective fact of the matter concerning which observable is measured by which measurement device.

There is one sense in which it is trivially true that, for some given measurement device, there may not be any such thing as *the* observable that it measures. Typically, any particular experimental device can be used for various different purposes and, accordingly, the quantity that is determined by the help of it depends on the specific context of application (and perhaps also on the way of evaluation of the data obtained through it). The quantum Bayesian claim that the measured observable is not an objective feature of the measuring device, however, goes far beyond these well-known observations as it includes the statement that, for any numerical value that has been obtained by means of an experiment, there is no fact of the matter concerning which observable it is a value *of*. For, if there is any fact of the matter concerning which observable some value d obtained through experiment is a value of, then, in the scenario mentioned by Fuchs, agents having registered the value d will have to assign the post-measurement state $\rho_d = \Pi_d$, which – by Fuchs' line of thought – 'must be an objective feature of the quantum system'.

This conclusion may appear unavoidable, but it is so hard to accept that it would not seem far-fetched to regard it, if it were valid, as a *reductio* of the whole project of giving an epistemic account of states. Agreement (at least to some high degree of approximation) on which observables the numerical values obtained in experiment are values *of* is pervasive among competent experimentalists. If no one could ever know to which observable some given value really belongs, it would quite generally be impossible to know the value of any observable. Since it is hard to deny that we often do have knowledge of the values of at least some observables, this consequence of Fuchs' reasoning is extremely problematic. Even if one adopts the quite radical view that the observables of microscopic systems (whatever one counts as such) never have determinate values, one can hardly make the same claim for those (normally macroscopic) systems to which we have more direct access and which are also treated quantum mechanically by means of the many-particle methods of quantum statistical mechanics (e.g. when computing heat capacities, magnetic susceptibilities, and the like). Here it is usually assumed that one has at least approximate knowledge of the volume, particle number, temperature, and maybe pressure of the (macro-) system, which gives constraints on the values of observables of the individual (micro-) particles.

However, the conclusion that we cannot obtain any knowledge of the values of observables seems contrived even with respect to microscopic systems that are not part of a many-particle macrosystem. If we consider, for instance, a Stern–Gerlach setup that measures S_x, the x-component of electron spin, it seems implausible to hold that there is no instant during the whole measurement process with respect to which an experimentalist can have knowledge of the value of S_x, which would be the case if there really were no fact of the matter as to which observable she has measured. When we look at how physicists actually talk and behave, we have, as it seems, strong evidence that obtaining knowledge of the values of observables is possible. I am not thereby saying that quantum Bayesianism owes us an 'ontological' account of under which conditions which observables do have determinate values, for this, as argued before, is something that can hardly be expected from an epistemic account of states that is not in terms of hidden variables. Something much more modest is asked for, namely that the mere possibility of having knowledge of the values of observables should not be ruled out as a matter of principle.

A final drawback of the view that the question of which observable is measured in which setup has no determinate answer is that it undermines the notion of a state assignment being performed *correctly*, which is also essential for quantum mechanical practice. In the case of systems having been prepared by a (so-called) *state preparation device*, for example, any state assignment that deviates from a highly specific one can reasonably be counted as wrong. Since state preparation can be regarded as a form of measurement, allowing the question of which observable is measured in which setup to have a determinate answer is the same as acknowledging the notion of a correct (or incorrect) state assignment. Quantum Bayesianism, arguably, should find a way of making sense of that notion.

There is a strategy to which defenders of quantum Bayesianism might try to resort in order to defend their account against the objection just presented, namely by trying to offer a story that explains why physicists may *in practice* behave as if the observable measured were determinate although, in reality, this is not so. They might try model such a story on the argument they have developed to account for the pragmatic success of physicists' talk of 'unknown quantum states', based on the quantum de Finetti theorem mentioned above.

As discussed, this argument impressively shows how talk about 'unknown quantum states' can be given an interpretation according to which it does not involve any commitment to quantum states as descriptions of quantum systems. However, there is no analogous way to interpret (and thereby justify) talk about 'the observable *A* measured by a certain device *D*' avoiding commitment to the definiteness of observables measured. There is at least one highly important difference between the two cases, namely that while we update our state assignment after having registered a measurement result, there is no prescription for adjusting our beliefs about the observable measured. So, there can be no derivation of a de Finetti-type reconstruction of talk involving 'the observable measured' along similar lines to the one involving talk about 'unknown quantum states'. It should also be noted that agreement on which observable has been measured has to be presupposed in the quantum Bayesian take on 'unknown quantum states': measurement results could never 'dictate' (Caves et al. [2002], p. 4541; see the last paragraph of the previous section for the full quote) that some specific state be assigned after a certain number of measurement trials unless we assume that in each case the observable measured is an objective matter. The assumption that different observers agree on which observable is measured in which case crucially enters the reasoning offered by Caves, Fuchs and Schack, but we are given no explanation of how this agreement is grounded.

What is needed to do better than quantum Bayesianism as an epistemic account of states is a justification of talk about measurement results 'dictating' state assignments without thereby reintroducing the notion of a state the system is in. In order to see whether such an account can be given, we must re-examine the claim made by Fuchs that in an epistemic account of states there can be no determinate answer to the question of which observable is measured by which setup.

5.3 Objectivity of observables measured in an epistemic account of states

According to Fuchs, there can be no determinate answer to the question of which observable is measured in which experimental setup, because if the observable measured were an objective feature of the device, the measured result would impose objective constraints on the state that has to be assigned to the system after measurement. Fuchs sees a conflict between this conclusion and the basic assumption of the epistemic conception of states that there is no agent-independent *true* state of the system. However, as I shall try to show now, this line of reasoning is not cogent. The idea of combining an epistemic account of states with the view that to a given setup there corresponds a determinate observable that is measured in that setup is perfectly coherent.

To see why this is the case, let us assume, in accordance with the epistemic conception of states, that the state one has to assign to a system should somehow depend on the information one has about the system. To this we now add the assumption that the measured observable is an objective feature of the setup and that agents registering measurement data may obtain information about the value of the observable that is measured.

Now, does it really follow from these assumptions that there is a certain state the system is in after measurement – the *true* state of the system, as one might call it? Clearly no: all that follows is that agents having registered a result must update the states they assign in a way that depends on the observable measured together with the registered result. As regards the case described by Fuchs, those having registered the 'click d', according to the assumptions made, are obliged to assign Π_d as their post-measurement state in order to perform their state assignments correctly. This, however, does not mean that Π_d must be regarded as the *true* or *real* post-measurement state of the system, the one it really *is in* after measurement. To reach this further conclusion, something more would have to be shown, namely that assigning a state that is different from Π_d would be wrong irrespective of what one may know of the system, i.e. wrong not only for those who know about the 'click d', but also for others who don't.

So let us consider more closely the situation of agents who are assigning states to the system – normally different from Π_d – without having had a chance to register the 'click d'. This may even be a matter of physical impossibility, due to the fact that the measurement process (or event) resulting in d lies outside their present backward light cone. Now, if we take seriously the idea that the states these agents assign reflect their epistemic situations with respect to the system, does it make sense to claim that, nevertheless, they are wrong to assign the states they assign, which are different from Π_d? Clearly not: their epistemic relations to the system, by hypothesis, are such that they have no reason at all for assigning Π_d. If their state assignments should indeed be adequate to their epistemic relations to the system, assigning Π_d would not only not be mandatory for them; it would even be *wrong*, for it would not conform to the knowledge they have of the values of observables of the system.

Assuming that the observable measured is an objective feature of the device, as we see, does not lead to the conclusion that there is a uniquely distinguished post-measurement state that must be assigned by anyone who intends to assign correctly. Consequently, definiteness of the observable measured does not imply that there is an agent-independent *true* state the system is in after measurement. We can therefore conclude that the argument given by Fuchs falls short of establishing that there can be no fact of the matter as to which observable is measured in which case if one assumes the epistemic conception of states.

It might be objected against this line of thought that it accords too much importance to the states assigned by those who are simply not well-informed about recent measurement results concerning a system. There are cases when measurement outcomes narrow down possible state assignments to a unique *pure* state, and this state, so the objection might go, clearly has a special status. Assignment of it is based on the best possible knowledge of the values of observables to be had and refusing to call it the state the system really is in may therefore seem artificial. Why should one accord any significance to the state assignments of those whose epistemic situation with respect to the system is simply worse?

To answer this challenge, it should first be emphasised that the epistemic conception of states has no problem admitting that knowledge of the values of observables can be better or worse (in the sense of being, say, more or less detailed, more or less precise, etc.) and that a state assignment based on excellent knowledge of the values of observables is likely to lead to the best predictive results. In that sense, when circumstances are such that an agent's knowledge of the values of observables of a quantum system cannot be improved further (at least not without losing part of this knowledge by measuring the system again) and narrows down possible states to assign to a unique pure state, there is nothing wrong with regarding this state as enjoying a special and privileged status. If one is careful not to jump to premature conclusions, one may – in those special circumstances – even call it *the* quantum state that quantum system is in.

It does not follow from conceding this much that, as proponents of ontic accounts of quantum states assume, for *any* quantum system in arbitrary circumstances there exists some state it is in which *would have to be assigned* by anyone assigning a state to the system if he or she were to assign correctly. In particular, systems that are subjected to measurement must in general be treated together with the measuring apparatus by assigning a combined system state that does not factorise into product states. Accordingly, systems on which measurements are carried out are not among those for which our knowledge of the values of their observables narrows down possible states to assign to a unique pure state. Thus, conceding that there exist conditions in which there is some quantum state which has a special and privileged state for the system does not reintroduce the measurement problem in the Maudlinian formulation as discussed in Chapter 3.

To conclude, it can consistently be allowed in an epistemic account of quantum states that the question of which observable is measured in which setup may have a determinate answer. One does not have to hold that knowledge of the values of observables is impossible in principle, as follows from the reasoning given by Fuchs.

Once we accept the view that the question of which observable is measured in which setup has a determinate answer, we are in a position, unlike the proponents of quantum Bayesianism, to make sense of the notion of a state assignment being performed correctly. From the perspective of the standard conception of states where states are seen as descriptions of quantum systems this may seem puzzling: according to this conception, a state assignment is correct if and only if it is an assignment of the state the system really is in. However, as I shall show in the following section, saying that a state assignment has been performed correctly remains coherent even if one rejects the notion of a state the system is in.

5.4 Constitutive rules

In order to preserve the notion of a state assignment being performed correctly without relying on the notion of a state a quantum system is in, we have to appeal to the rules employed in the assignment of quantum states, arguing that assigning correctly means to assign in accordance with them. Assuming the epistemic conception of states to be valid, we should think of these rules as determining the state an agent has to assign to the system, depending on her knowledge of the values of its observables. Let me briefly discuss the most basic examples of such rules by going through the different contexts of application in which they are relevant.

One type of application of quantum theory that needs to be considered is one where a system is considered over a time interval and left undisturbed by external influences which might be exploited to gain new knowledge about the values of its observables. In these cases, the state of the system must be evolved in time following unitary time-evolution as determined from the time-dependent Schrödinger equation. What is normally seen as a *fact* about quantum states – that their time-evolution follows the Schrödinger equation – takes the form of a rule of state assignment, namely the rule that an agent should apply unitary time-evolution for all times t with respect to which she has no new incoming data concerning the values of observables of the system.

Another prescription has to be used in case the evolution of the system is affected in such a way that it does become possible to extract information about the values of its observables – in the simplest case, that the value of an observable A lies within a certain range Δ. This is Lüders' Rule, the generalisation of von Neumann's projection postulate Eq. (2.16) to the situation where the quantum state is given as a density matrix. Lüders' Rule can be motivated as the analogue of Bayes' Rule for probability conditionalisation in the light of new evidence in a non-commutative setting.[35] If we denote by ρ the state assigned to the system immediately before the measurement occurs and by Π_Δ the projection onto the linear span of eigenvectors of A with eigenvalues lying within Δ, the change of state according to Lüders' Rule is given by

$$\rho \longrightarrow \rho_\Delta = \frac{\Pi_\Delta \rho \Pi_\Delta}{\mathrm{Tr}(\Pi_\Delta \rho \Pi_\Delta)}. \tag{5.1}$$

Both unitary time-evolution in accordance with the Schrödinger equation and collapse in accordance with Lüders' Rule are rules of state change. They can be applied to the state of the system only if a state has already been assigned in the first place. We might, however, also be interested in the standard of correctness for the assignment of states to systems where no state has been assigned before. In some cases this problem can be solved by appeal to Lüders' Rule alone, namely when one's knowledge of the values of observables uniquely fixes the post-measurement state so that the pre-measurement state – if one had been assigned – would not have any influence on the post-measurement state.

In the generic case, however, this is not sufficient for determining the state that one has to assign. In quantum statistical mechanics, for example, one is usually dealing with systems for which one has knowledge of only a very limited number of quantities, typically called 'macroscopic variables' such as temperature, pressure, magnetisation, etc. Here one expects that the state to be assigned should conform to the criterion that it maximises entropy subject to certain constraints that are determined from what one knows of the values of these variables.[36] Entropy maximisation, as emphasised by Jaynes, can be motivated on the grounds that it leads to 'the only unbiased assignment [of probabilities] we can make; to use any other would amount to arbitrary assumption [*sic*] of information which by hypothesis we do not have' (Jaynes [1957a], p. 623). In quantum mechanics, entropy must be given as a function of the state assigned to the system, and it is widely believed that the von Neumann entropy $S(\rho) = -k_B \text{Tr}(\rho \log(\rho))$ is the appropriate quantity here. Although there has been some debate on whether this view is really correct,[37] we need not concern ourselves with this question here. It is sufficient for us to assume that an entropy function exists that has to be maximised when a state is assigned to a system to which no state has been assigned before.

Applications of quantum theory are directed at various different aims among which predicting the outcomes of future measurements and the associated statistical data is just one.[38] Others include explaining the observed physical behaviour of physical systems (an aim which Chapter 9 discusses in more detail) and determining a quantum system's internal structure and the interactions among its constituents. To achieve these latter aims, one typically needs to *start* with the statistical data, *reconstruct* a quantum state to be assigned from these data, and then to conclude what the system's internal structure and the interactions among its constituents are, such that knowledge of them, given the rules of state assignment already mentioned, would require assigning the reconstructed quantum state. In this procedure, one needs, first, a 'meta-rule' of quantum state assignment, namely that identical quantum states must be assigned to identically prepared quantum systems – where what counts as 'identical preparation' includes 'identical knowledge of the values of observables' – which rule plays the same role as the assumption of 'exchangeability' in the quantum Bayesian account of quantum state tomography sketched above. Second, one needs the Born Rule (Eq. (2.6)) – usually thought of as an instrument to extract probabilities from quantum states – which is here employed in the inverse direction: as a rule for reconstructing a quantum state from statistical data.

The *status* of the rules governing state assignment just mentioned in the present epistemic account of quantum states deserves special mention. It is, in particular, very different from the status these principles have in ontic accounts of quantum states. The nature of this difference can be clarified by bringing into play some terminology invented and introduced by John Searle in the context of his theory of speech acts in (Searle [1969]). Searle distinguishes between two different sorts of rules, and this distinction is very useful for clarifying the role of the rules of state assignment in the epistemic account of states proposed here.

The distinction between the two kinds of rules is introduced by Searle as follows:

> I want to clarify a distinction between two different sorts of rules, which I shall call *regulative* and *constitutive* rules. I am fairly confident about the distinction, but do not find it easy to clarify. As a start, we might say that regulative rules regulate antecedently or independently existing forms of behavior; for example, many rules of etiquette regulate inter-personal relationships which exist independently of the rules. But constitutive rules do not merely regulate, they create or define new forms of behavior. The rules of football or chess, for example, do not merely regulate playing football or chess, but as it were they create the very possibility of playing such games. The activities of playing football or chess are constituted by acting in accordance with (at least a large subset of) the appropriate rules. Regulative rules regulate pre-existing activity, an activity whose existence is logically independent of the rules. Constitutive rules constitute (and also regulate) an activity the existence of which is logically dependent on the rules. (Searle [1969], pp. 33f.)

According to the standard view of quantum states as states quantum systems 'are in', the rules of state assignment are regulative rules. To see this, assume that a system is indeed correctly described by an agent-independent true state. The existence of such a state, according to this perspective, is independent of whether there are any agents who might actually happen to assign it. Therefore, what agents are aiming at when assigning a state – namely, to assign the state the system really is in – can be specified without mentioning the rules one follows in order to achieve this goal. Consequently, according to the standard view of quantum states as representing physical features of quantum systems, the notion of a state assignment being performed correctly is 'logically independent of the rules' according to which it is done, namely those discussed before in this section. If states are seen as describing the properties of quantum objects, the rules of state assignment play the role of an instrument or a guide that is used to arrive at the true state of the system (or some reasonable approximation to it). State assignment, from this point of view, is a 'form of behaviour' that makes sense and therefore exists 'antecedently [to] or independently' of the rules according to which it is done. It is *regulated* by these rules without being *constituted* by them, in the sense these expressions have in the writings of Searle.

In a version of the epistemic conception of states that tries to preserve the notion of a state assignment being performed correctly, however, the status of the rules of state assignment must be different: their role cannot be that of a guide or instrument to assign a quantum state that is hoped to be (a good approximation of) the *true* one of the system, for the notion of such a quantum state is rejected. Rather, assigning a state in accordance with the rules of state assignment is what it *means* to perform a state assignment correctly, so the notion of a state assignment being performed correctly will have to be *defined* in terms of these rules. Consequently, the notion of a state assignment being performed correctly is, to use Searle's expression, 'logically dependent on the rules' according to which it is done. We therefore have to conceive of the rules of state assignment as constitutive rules if we want to preserve the notion of a state assignment being performed correctly without being forced to accept the notion of a state a quantum system is in. Due to the eminent role this account attributes to the rules governing state assignment, I propose to call it the 'Rule Perspective'. The next section discusses what quantum probabilities should be claimed to be probabilities *of* in the Rule Perspective and what *kinds* of probabilities quantum probabilities are according to the Rule Perspective.

6
Quantum Probabilities: What Are They?

The previous chapter addressed the challenge of making sense of the notion of a quantum state assignment being performed correctly without thereby acknowledging the notion of a quantum state a quantum system 'is in'. The present chapter discusses what interpretation of quantum probabilities goes best with the account that I have called the 'Rule Perspective' and started to develop in the previous chapter. The first section considers what quantum probabilities should be taken to be probabilities *of* in the Rule Perspective; the second investigates what philosophical interpretation of probabilities applies best to quantum probabilities thus conceived. The third section considers two objections to the resulting account of quantum probabilities.

6.1 Probabilities of what?

An account of quantum probabilities that goes very well with the account of the rules of state assignment just outlined has recently been proposed by Richard Healey in the context of his pragmatist interpretation of quantum theory (Healey [2012a]). A central notion in that interpretation is that of a *non-quantum magnitude claim* ('NQMC'), which refers to a statement of the form 'The value of observable A of system s lies in the range Δ' or '$v(A) \in \Delta$' for short. Healey calls such statements 'non-quantum', arguing that 'NQMCs were frequently and correctly made before the development of quantum theory and continue to be made after its widespread acceptance, which is why I call them non-quantum' (Healey [2012a], p. 760). NQMCs more or less correspond to what for adherents of the Copenhagen interpretation were descriptions in terms of 'classical concepts',[39] but Healey objects to this usage of 'classical' as inviting the misleading impression that a NQMC 'carries with it the full content of classical physics' (Healey [2012a], p. 740). An endorsement of a NQMC is not to be construed as entailing any commitment to the view that the dynamics of the system at issue are described by classical laws of motion. In Healey's view, taken over in the Rule Perspective, NQMCs are crucial in the application of quantum theory in that they (not quantum states) have the linguistic function of describing the phenomena and regularities quantum theory is used to predict and explain.

NQMCs are naturally regarded as the bearers of quantum probabilities, as can directly be seen from the Born Rule (2.6), which, to recapitulate, is given by

$$\Pr_\rho(v(A) \in \Delta) = \mathrm{Tr}(\rho \, \Pi_\Delta^A), \tag{6.1}$$

where ρ denotes the density operator assigned to the system, $v(A)$, the value of A, and Π_Δ^A, the projection on the span of eigenvectors of A with eigenvalues lying in Δ. It is natural to read this equality as attributing a probability to a statement of the form 'The value of A lies in Δ', that is, to a NQMC.

The Rule Perspective reads Born Rule probability ascriptions to NQMCs in this straightforward way, but it has to acknowledge the no-go theorems due to Gleason, Bell, Kochen, Specker, and others, which have to be considered when attributing sharp values to the observables of a quantum system (see Appendices A and B). A standard response to the difficulties brought to the fore by these theorems is to interpret the Born Rule as attributing probabilities to NQMCs only inasmuch as they concern possible *measurement outcomes*. According to this view, the quantity $\mathrm{Tr}(\rho \Pi_\Delta^A)$ is interpreted as the probability of obtaining a value of A lying in Δ if A were to be *measured*. This (instrumentalist) take on the Born Rule has the unappealing feature that it construes the empirical relevance of quantum theory as restricted to measurement contexts. In addition, it seems not to do justice to quantum theoretical practice, where claims about the values of observables are often considered (and Born Rule probabilities computed for them) even where these values are not 'measured' or otherwise determined experimentally.[40] This observation makes the challenge of clarifying the appropriate range of applicability of the Born Rule even more pressing.

The response to this challenge found in the literature which seems to go best with the Rule Perspective is essentially that of Healey's pragmatist interpretation of quantum theory. Appealing to environment-induced decoherence, it holds that Born Rule probabilities apply precisely to those NQMCs which are in terms of observables for which taking into account the system's interaction with its environment renders the density operator assigned to the system (at least approximately) diagonal. In Healey's own words:

> Born-rule probabilities are well-defined only over claims licensed by quantum theory. According to the quantum theory, interaction of a system with its environment typically induces decoherence in such a way as (approximately) to select a preferred basis of states in the system's Hilbert space. Quantum theory will fully license claims about the real value only of a dynamical variable represented by an operator that is diagonal in a preferred basis: it will grant a slightly less complete license to claims about approximately diagonal observables. All these dynamical variables can consistently be assigned simultaneous real values distributed in accordance with the Born probabilities. So there is no need to formulate the Born rule so that its probabilities concern only measurement outcomes. (Healey [2012a], p. 749)

To a first approximation, I propose to read these remarks as follows: a NQMC of the form 'The value of A lies in Δ' is 'licensed' by quantum theory just in case an agent who applies quantum theory to the system in question is entitled to assume as the basis of her further reasoning that the value of A either determinately lies within Δ or outside Δ. In accordance with Healey's claim quoted above the Rule Perspective should say that quantum theory 'licenses' those NQMCs which ascribe values to observables A for which the (reduced) density matrix assigned to the system is at least approximately diagonal in a preferred way by the spectral decomposition of A.[41] The Rule Perspective concurs with Healey's pragmatist interpretation in that it construes Born Rule probabilities as 'well-defined only over claims licensed by quantum theory' in precisely that sense.

Environment-induced decoherence is relevant here in that taking into account the system's coupling to its environment and performing the trace over the environmental degrees of freedom typically makes the reduced density matrix (at least approximately) diagonal in an environment-selected basis. In conditions which function as measurement setups the role of the environment is typically played by the system which is used as the measurement apparatus, and the measurement fulfils its purposes if an eigenbasis of the observable meant to be 'measured' coincides with the Hilbert space basis selected by decoherence. This accounts for the fact that quantum theory, when employed with competence, licenses application of the Born Rule in measurement contexts and thus yields probabilities for the possible values of the observable(s) the apparatus is meant to measure.

Healey himself elucidates the expression of a NQMC being 'licensed' by resorting to an inferentialist account of semantic content that regards the content of a NQMC as determined by what 'material inferences'[42] an agent applying quantum theory is entitled to draw from it. An example of a material inference is from 'Lightning is seen now' to 'Thunder will be heard soon' (Brandom [1994], p. 98). Material inferences are not logically gapless, but nevertheless crucial and abundant in our everyday practical (and scientific) reasoning.

To provide an example of a material inference that is *not* legitimate in quantum theory Healey appeals to the standard double-slit setup with a characteristic interference pattern on a screen behind the double-slit. Here the inference from 'The particle passes either through the upper or lower slit' to 'It is possible reliably to observe through which slit each particle passed without altering the interference pattern' (Healey [2012a], p. 746), natural though it seems, is illegitimate. Accordingly, the NQMC 'The particle passes either through the upper or through the lower slit' is not 'licensed' (though it may be 'considered'), and this is due to the fact that the particles are not subject to environment-induced decoherence when passing through the setup. The density matrices to be ascribed to them are not approximately diagonal (and not even appropriately block-diagonal) in the position basis, and application of the Born Rule to determine probabilities for whether a particle passes through the upper or lower slit is illegitimate.

By the standards of the inferentialist account of semantic content Healey appeals to, only those NQMCs that are licensed have well-defined semantic content at all. Thus, the more important the effect of environment-induced decoherence gets in a setup, the more clearly shaped the content of the NQMCs attributing sharp values to the observables 'measured' gets. Healey warns his readers that this 'progressive definition of content has no natural limit such that one could say that, when this limit is reached, a statement like ["The particle passes either through the upper or lower slit"] is simply true because one has finally succeeded in establishing a kind of natural language-world correspondence relation in virtue of which the statement correctly represents some radically mind- and language-independent state of affairs' (Healey [2012a], p. 747). Thus, that a NQMC is 'licensed' for an agent does not give one any guarantee that it is determinately true (or false, as one might add). All that it means, according to Healey, is that it has well-defined semantic content. Is it plausible that claims of the form 'The value of A lies in Δ' have no clearly shaped content unless they are 'licensed' and the Born Rule applies to them due to how decoherence effects appear in the quantum state assigned? And is it plausible that, while 'The value of A lies in Δ_1' may not have any well-defined content, 'The value of A lies in Δ_2' for some other interval Δ_2 may well do, depending on how decoherence acts on the system? Answering 'no' to these questions would not mean rejecting Healey's (arguably very plausible) diagnosis that the Born Rule is legitimately applied only to those NQMCs for which decoherence effects make the density matrix (at least approximately) diagonal in an environment-selected basis. One may well acknowledge that applying the Born Rule to NQMCs that are not 'licensed' in this sense is illegitimate, but conclude that this tells us something about the Born Rule – namely, for which NQMCs it produces reliable probabilities – rather than about which NQMCs have well-defined content. I find this reaction attractive and, in Section 12.2, will explore whether the Rule Perspective is compatible with an especially austere version of the idea that all NQMCs have determinate content, namely, the idea that all observables simultaneously have sharp values for each quantum system. In the meantime, however, I would like to leave it open whether being 'licensed' is relevant to a NQMC's degree of content. Quantum theoretical practice, at least, does not seem to dictate any particular answer to this question.

6.2 Probabilities of what kind?

Characterising the status of quantum probabilities in the Rule Perspective in terms of the subjective/objective dichotomy is not completely straightforward. On the one hand, quantum probabilities are 'subjective' according to the Rule Perspective in that different agents ('subjects') should ascribe different probabilities to one and the same NQMC if the agents' epistemic conditions differ. On the other hand, they are 'objective' in that for sufficiently specified epistemic conditions the probability to be ascribed to a NQMC is completely fixed by the rules governing state assignment, including the Born Rule. In other words, quantum probabilities are *non-objective* inasmuch as they are not *invariant* under changes in the epistemic situations of the agents who ascribe them; they are *objective* inasmuch as they are fixed as soon as all relevant features of the epistemic conditions of the agents ascribing them are made *explicit*.[43]

To conclude, the Rule Perspective interprets quantum probabilities as objective in important respects. Can this sense of objectivity be captured by any of the most popular objectivist conceptions of probabilities? To be specific, is the Rule Perspective compatible with a frequentist or propensity interpretation of quantum probabilities?

6.2.1 Relative frequencies?

Interpretations of probability as relative frequency are falling on hard times. Hardly anyone doubts that probability and relative frequency are somehow connected, for example in that data about relative frequencies are used to test hypotheses about probabilities, but there seems to be widespread agreement among philosophers today that one cannot simply identify probabilities with relative frequencies. If one does so and conceives of the relevant collectives as actual and finite ('actual frequentism'), it follows that, implausibly, probabilities cannot possibly fail to agree exactly with the actual relative frequencies within these collectives. Actual frequentism, by itself, is quite clearly inadequate to account for how we actually use the concept of probability, in particular also in quantum theory. Nevertheless, actual frequencies may be fruitfully connected to the quantum probabilities in a more indirect way than envisaged by actual frequentism. I explore this idea in the last chapter of this work.

In response to the difficulties with which actual frequentism is confronted, frequentists often resort to an idealisation of the relevant collectives and take them to be merely hypothetical and infinite. In that case, however, other serious problems of a general character arise. As explained by Alan Hájek and Ned Hall (Hájek [2009] gives a devastating, much more detailed criticism):

> [W]e sometimes attribute non-trivial probabilities to results of experiments that occur only once. The move to hypothetical infinite frequencies creates its own problems: There is apparently no fact of the matter as to what such a hypothetical sequence would be, not even what its limiting frequency for a given attribute would be, nor even whether that limit is even defined; and the limiting relative frequency can be changed to any value one wants by suitably permuting the order of trials. (Hájek and Hall [2002], p. 161)

Due to all these widely acknowledged difficulties for hypothetical frequentism, the idea of combining the Rule Perspective with hypothetical frequentism about quantum probability does not seem very tempting

from the start. However, one might speculate that there is on independent grounds a profound incompatibility between the Rule Perspective
and hypothetical frequentism due to the fact that the latter, contrary
to the Rule Perspective, is a thoroughgoing *objectivist* account of probability: probabilities, according to frequentism in any of its forms, are
objective in that what the relative frequencies *are* in some (completely
ordered) collective is an objective matter that does not depend on
any agent. However, the fact that quantum probabilities, when conceived of as relative frequencies, are objective in *this* sense would not be
problematic if one wanted to add frequentism to the Rule Perspective.

To see this, consider, for the sake of specificity, what seems to be the
best-known account of quantum probabilities as hypothetical infinite
relative frequencies found in the literature, namely, Leslie E. Ballentine's *statistical interpretation* of quantum theory.[44] In accordance with
the spirit of frequentism about probabilities, it is centred around 'the
assertion that a quantum state (pure or otherwise) represents an *ensemble* of similarly prepared systems' (Ballentine [1970], pp. 360–361). For
Ballentine, the identity of an ensemble of systems in the relevant sense
is constituted by a common preparation method:

> For example, the system may be a single electron. Then the ensemble
> will be the conceptual (infinite) set of all single electrons which have
> been subjected to some state preparation technique (to be specified
> for each state), generally by interaction with a suitable apparatus.
> (Ballentine [1970], p. 361)

The actual relative frequencies in a finite run of experiments, according
to Ballentine, are mere *approximations* of the real probabilities, and these
approximations are exact only in the infinite limit:

> The probabilities are properties of the state preparation method and
> are logically independent of the subsequent measurement, although
> the statistical frequencies of a long sequence of similar measurements
> (each preceded by state preparation) may be expected to approximate
> the probability distribution. (Ballentine [1970], p. 361)

It may seem as if Ballentine's view, by conceiving of the probabilities as
objective 'properties of the state preparation method', were fundamentally incompatible with accounts such as the Rule Perspective which
conceive of quantum probabilities as closely connected with the epistemic conditions of those who ascribe them. Ballentine himself briskly
dismisses what he terms 'the frankly subjective interpretation […],

according to which the quantum state description is not supposed to express the properties of a physical system or ensemble of systems but our knowledge of these properties' (Ballentine [1970], p. 371). The association of quantum probabilities with what the agents know about the system, as he argues, is 'not so much [...] wrong as [...] irrelevant' (Ballentine [1970], p. 371). Since the Rule Perspective is an epistemic account of quantum states, it might seem as if frequentism, Ballentinean frequentism at least, were fundamentally incompatible with it.

This impression is misleading, however, for the Rule Perspective differs crucially from the 'frankly subjective interpretation' Ballentine feels free to dismiss so easily. While it does emphasise a relation between quantum state assignments and the assigning agents' epistemic conditions, it does not conceive of the quantum states (and probabilities) as *expressing* (or describing) these epistemic conditions. Inasmuch as the quantum state 'expresses' anything at all, it expresses something about the system it is assigned to in that it depends on the values of its observables – inasmuch as they are known by the agent. But this means that the locus of objectivity in the assignment of quantum states is very similar according to the Rule Perspective and according to Ballentine's statistical interpretation: in Ballentine's account, quantum probabilities are objectively determined by the preparation device used, in the Rule Perspective they are objectively determined by the epistemic conditions of those who ascribe them, and these in turn may depend on the preparation method used to create quantum systems with specific desired properties.

To conclude, the move of conceiving of a probability ascription as an attribution of some hypothetical infinite collective (or 'ensemble') can be made in the Rule Perspective no less than in Ballentine's interpretation. Inasmuch as what the relative frequencies are within some collective is an objective matter in Ballentine's account, it is equally objective in the Rule Perspective supplemented with talk about hypothetical infinite collectives. So, even if one ignores all the disadvantages of hypothetical infinite frequentism mentioned above, resorting to it does not (by itself) result in an account of quantum probabilities that accords them a significantly more objective status than the Rule Perspective.

6.2.2 Propensities?

Objective probabilities in the sense of objective *tendencies* or *dispositions* for 'coming into being' are referred to as *propensities* in the literature.[45] Different types of propensity interpretations of probability were proposed, e.g., by Popper, Giere, and Gillies, and early ideas about probabilities as propensities go back at least to C. S. Peirce. Many contemporary writers argue that, among all interpretations of probability, propensity interpretations apply best to quantum probabilities.[46] In fact, authors who endorse pluralistic accounts of 'probability', according to which many different conceptions of probability are legitimate, quantum probabilities are sometimes named as the paradigm instantiations of probabilities as propensities.[47] Popper himself claims that considerations on the quantum mechanical two-slit experiment stimulated his propensity interpretation:

> [T]he interpretation of the two-slit experiment ... convinced me that probabilities must be 'physically real' – that they must be physical propensities, abstract relational properties of the physical situation, like Newtonian forces, and 'real', not only in the sense that they could influence the experimental results, but also in the sense that they could, under certain circumstances (coherence), interfere, i.e. interact, with one another. (Popper [1959], p. 28)

From the point of view of the Rule Perspective, however, the importance of the concept of propensity probability in the interpretation of quantum theory lies mostly in the fact that it illustrates what quantum probabilities are *not*. A propensity, recall, is an objective property of a physical situation (which may include the complete configuration of physical properties at a time), best conceived of as a tendency to develop in a certain way into a different situation at a later time. The propensities, on this view, are what *governs* the evolution of reality, and the different propensities that are present in a given physical situation may well pull the properties of the systems in different directions.

But this way of locating the objectivity of probabilities, somehow *inside* the physical systems (or situations) and somehow 'steering' them through time is exactly the perspective on quantum probabilities which the Rule Perspective discourages. For according to the Rule Perspective quantum probabilities are objective inasmuch as there are objective standards of correctness for their assignment, not in that they are somehow present as physical quantities *within* the systems. The difference is crucial and profound. I will argue later (in Chapter 10) that the alleged tension between quantum non-locality and special relativity arises largely from an inadequate picture of quantum probabilities as propensities which govern the evolution of quantum systems through time.

6.2.3 Objective probabilities as constraints on rational credences

What are the features in virtue of which some quantity x that appears in a scientific theory qualifies as an objective probability? The trivial (and not yet very informative) answer to this question is that it must in some way fulfil the characteristic *role* of an objective probability. Wherein lies that specific role? A widely acknowledged answer to this question is given by David Lewis, who recommends that one should not 'call any alleged feature of reality "chance" unless [one has] already shown that [one has] something, knowledge of which could constrain rational credence' (Lewis [1994], p. 484). The idea expressed here nicely sums up the core of his 'subjectivist guide' to objective probability[48] ('chance'), which finds its more formal expression in a prescription that he calls the *'Principal Principle'*. Intuitively, what that principle says is that, in order to be rational, an agent's credences should coincide with the objective probabilities, provided the agent has epistemic access to these. Some of the details that arise in the formulation of the Principal Principle (in particular the role of the notion of *admissible evidence* in Lewis' formulation of it) are discussed in detail in Sections 10.3 and 10.4, where its ramifications for Bell's notion of local causality are investigated.

The Principal Principle leaves room for various philosophical interpretations of objective probabilities such as frequentism and propensity interpretations, provided the quantities singled out as probabilities by these interpretations can reasonably be claimed to appropriately impose constraints on the rational credences of (actual or hypothetical) agents. Lewis himself offers a 'best-system' analysis of objective probabilities that regards the facts about them as depending in a potentially non-trivial way on the symmetries and frequencies in the distribution of properties across space-time. As I argue in detail in Chapter 12, the 'best-system' approach to objective probabilities is arguably very compatible with the Rule Perspective and can help us understand how quantum theory provides insights into physical reality even if the quantum states themselves do not correspond to anything real.

Independently of whether one accepts the 'best-system' analysis of objective probabilities one should acknowledge the Principal Principle as a useful means of characterising what it takes for a quantity to be an objective probability. In particular, it is arguably relevant to the quantum probabilities derived from legitimate application of the Born Rule: that some quantity of the form $\mathrm{Tr}(\rho \Pi_\Delta^A)$ is adequately regarded as a probability is due to the fact that it gives the rational credence with respect to the NQMC 'The value of A lies in Δ' for an agent who assigns the state ρ to the system, provided the NQMC is 'licensed' in the Healeyan sense.

In the following section I discuss two objections to the perspective on quantum probabilities outlined in this section.

6.3 Objections to this interpretation of probabilities

6.3.1 The means/ends objection

Interpretations which regard quantum probabilities as in some way depending on (or tailored to) the epistemic situations of the subjects who ascribe them have been criticised on a number of grounds. Since the Rule Perspective conceives of quantum probabilities as in some respects subjective (as outlined in the beginning of the previous section), it makes sense to consider two especially pointed (and closely related) criticisms of accounts of quantum probabilities as subjective and see whether they apply to the Rule Perspective. Since they pertain to the interpretation of quantum probabilities specifically, I discuss them in this chapter rather than together with the other objections in Part III of this book.

Both objections concerning quantum probabilities to be discussed now were originally formulated as objections against quantum Bayesianism by Chris Timpson. My conclusion will be that despite the important points of agreement between quantum Bayesianism and the Rule Perspective, they do not gain force against the latter.

The objection I consider first is referred to by Timpson as the *means/ends objection*. Its underlying idea is that any account which denies quantum probabilities the status of objective features of the world inevitably makes it mysterious how the theory helps us with as little as 'the pragmatic business of coping with the world', let alone with the more ambitious goal of 'finding out how the world is' (Timpson [2008], p. 606). According to Timpson, any interpretation of quantum probabilities that does not conceive of them as objective single-case probabilities makes it unclear why updating our assignments of probabilities in the light of new data should be useful and enhance our predictive success: 'if gathering data does not help us track the extent to which circumstances favour some event over another one (this is the denial of objective single case probability), then why does gathering data and updating our subjective probabilities help us do better in coping with the world [...]?' (Timpson [2008], p. 606) What makes

Timpson's worries most pressing is that for the quantum Bayesian the question of *in which way* an agent should update her probability assignments in the light of new data does not have an objective answer. The reason for this is that quantum Bayesianism does not acknowledge any fact of the matter as to which observable is being measured in which setup and what update of one's state assignment one should make. On this view, the predictive and pragmatic success of quantum theory – why it helps us in 'coping with the world' – is indeed mysterious: if there is no objective answer as to *how* some assignment of probabilities should be updated in the light of new evidence, it becomes a miracle that updating probabilities is of use at all. The main force of the means/ends objection to quantum Bayesianism, as we see, derives from the quantum Bayesian's claim that we can never know what observable some measured value is a value of.

Timpson outlines a possible quantum Bayesian reply to this challenge, which, as he says, is 'of broadly Darwinian stripe'. According to this reply, '[w]e just do look at data and we just do update our probabilities in light of it; and it is just a brute fact that those who do so do better in the world; and those who do not, do not' (Timpson [2008], p. 606). However, as he argues, this response is ultimately unsatisfying in that it does not address the original worry, namely, the nagging question 'why do those who observe and update do better' (Timpson [2008], p. 606). Given that, for quantum Bayesianism, no way of updating in the light of incoming data counts as correct (in contrast to the other ways one might think of), the challenge is particularly serious.

However, as a moment's reflection makes clear, this problem is not generic in epistemic accounts of quantum states. The Rule Perspective, for instance, avoids it. To see this, recall that according to the Rule Perspective quantum theory helps us determine and predict which *non-quantum* claims (NQMCs) are true. So, on this view the theory is not only a tool that helps us accomplish the 'business of coping with the world' but also one that helps us at least in some way with the more ambitious goal of 'finding out how the world is', to use Timpson's words (for more on in which sense it does so, see Chapter 12). In contrast to quantum Bayesianism, the Rule Perspective concedes that we often do have knowledge of the values of observables, and it regards the practice of making probability ascriptions and updating them in the light of new evidence as directed at the aim of improving that knowledge in various (direct and indirect) ways. The Rule Perspective does not deny that an intimate connection exists between objective features of the world, as described in terms of NQMCs, and quantum state assignments, performed on grounds of what NQMCs one knows to be true. Therefore, it is not miraculous that updating our state assignments – and with them our probability ascriptions – can help us predict which NQMCs are true. There remains no 'explanatory gap' (Timpson [2008], p. 606), as Timpson objects to quantum Bayesianism, between the methods of enquiry – assigning quantum states and deriving probabilities from them – and the goals we seek to achieve by applying quantum theory – broadly (and somewhat crudely), to determine which NQMCs are true. Furthermore, since knowledge about 'how the world is' is plausibly helpful for our competence in 'coping with the world', it is small wonder that quantum theory helps us with the latter if it helps us with the former.

6.3.2 The quantum Bayesian Moore's paradox

The second of Timpson's criticisms against the interpretation of quantum probabilities as subjective focuses on assignments of probability 1. According to Timpson, if one conceives of quantum probabilities as subjective degrees of belief, this commits one to the systematic endorsement of pragmatically problematic sentences of the 'quantum Bayesian Moore's paradox' type. Sentences of this type are cousins of the better-known 'Moore's paradox' sentences, invented by G. E. Moore, which are characterised by having the form

p, but I don't believe that p.

There is a long-standing philosophical debate on the status and proper interpretation of these sentences, in particular as to whether they involve a pragmatic or even a semantic contradiction, but there seems to be agreement on their paradoxical nature in that, as expressed by Timpson, they 'violate the rules for the speech act of sincere assertion' (Timpson [2008], p. 602). Timpson argues that by interpreting quantum probabilities as subjective degrees of belief and denying that there is such a thing as *the* quantum state a quantum system is in, quantum Bayesianism is committed to the systematic endorsement of sentences having a similar structure and a similar paradoxical flavour. The problem he diagnoses occurs in connection with the assignment of quantum states ascribing probability 1 to a possible value of at least one observable. Typical examples arise from the assignment of *pure* quantum states. These ascribe only probabilities 0 or 1 to the possible values of observables they are eigenstates of. The problem can be seen by considering an agent who consciously accepts the quantum Bayesian take on quantum probabilities as subjective degrees of belief and assigns a pure quantum state to a system, for instance the state $|\uparrow_z\rangle$ for the spin degree of freedom of a spin-1/2 system. Such an agent, according to Timpson, 'must be happy to assert sentences like: "I assign a pure state (e.g. $|\uparrow_z\rangle$) to this system, but there is no fact about what the state of this system is"' (Timpson [2008], p. 604). In other words, any quantum Bayesian agent is committed to the systematic endorsement of sentences of the form:

QBMP: 'I am certain that p (that the outcome will be spin-up in the z-direction) but it is not certain that p.' (Timpson [2008], p. 604. The acronym 'QBMP' stands for 'quantum Bayesian Moore's paradox'.)[49]

For a quantum Bayesian, an ascription of probability 1 to the value of an observable signals complete certainty as to what the outcome of measurement of that observable will be, but her subjective Bayesian take on probabilities (including probability 1) implies that she cannot claim that there is any *objective* certainty as to what the measurement outcome will be since she cannot countenance any 'fact determining what the real state is' (Timpson [2008], p. 605). Sentences of the form of the QBMP seem pragmatically problematic since expressing absolute certainty seems irrational if one does not believe it to be grounded in facts. Timpson notes that similar-structured paradoxical features of ascriptions of probability 1 (and, one might add, of ascriptions of probability 0) are generic in accounts that are based on the subjective Bayesian take on probability and arise not only in the context of quantum Bayesianism. However, whereas those who hold subjective Bayesian views with respect to other contexts are in principle free to refrain from making extremal probability assignments, quantum Bayesians must make them unless they are prepared to abandon the assignment of pure states. As Timpson notes, 'the occurrence of these paradoxical sentences isn't just an occasional oddity which can be ignored', but one which 'arises whenever one finds a quantum Bayesian who is happy to assign pure states and is also explicit about what their understanding of the quantum state is' (Timpson [2008], p. 605).

There is a further difficulty for quantum Bayesianism here that is not even mentioned by Timpson but seems no less severe. It arises again from the quantum Bayesians' denial that we can ever have any knowledge of the values of observables or, what comes to the same, that we can ever have any knowledge as to which NQMCs are true of some system. According to quantum Bayesianism, we can never know that some NQMC 'p' is true, not even in cases where we assign pure states, which, on the quantum Bayesian reading, invariably signal our being certain that p is true. It is difficult to see how one might coherently claim that agents cannot *know* that p, yet be *certain* that p.

Timpson considers a possible quantum Bayesian reaction to this problem, namely, to adopt a perspective on Born Rule probability ascriptions that is similar to that of ethical non-cognitivists on moral discourse. In analogy to how non-cognitivists may aspire to explain why endorsing moral claims can be rational and legitimate despite the claimed non-existence of moral facts, quantum Bayesians might 'elaborate on how there can be a role for personal certainty within our intellectual economy which is insulated against the absence of any impersonal certainty' (Timpson [2008], p. 606). The quantum Bayesian, as suggested by Timpson, may try to defend her position along similar lines against the challenge sketched in the previous paragraph by suggesting that one can indeed be coherently certain that p while at the same time – on some meta-level – regarding it as impossible to know that p. Even though this move might be viable in principle, it still appears rather far-fetched and contrived.

Unlike quantum Bayesianism, the Rule Perspective does not conceive of quantum probabilities as *expressing* the ascribing agents' degrees of belief. However, it acknowledges that the Born Rule probabilities one ascribes often do have ramifications for one's rational credences. Agents who accept the Rule Perspective and apply the Principal Principle in cases where the states they assign deliver Born Rule probability 1 for a NQMC 'p' will indeed be certain that p. It may therefore seem that the proponent of the Rule Perspective has the same problems as the quantum Bayesian with respect to QBMP-type sentences in that both are committed to the systematic endorsement of sentences of the form 'I am certain that p, but it is not certain that p'.

However, the Rule Perspective differs from quantum Bayesianism in that it acknowledges that whether a NQMC 'p' is true is something we can (under favourable circumstances) *know*. There is therefore no reason for an adherent of the Rule Perspective to deny that her assignment of probability 1 to some NQMC is related to a fact she has evidence of, namely, the fact that things are indeed as described by 'p'. Inasmuch as impersonal certainty is construed as entailed by truth (that is, if 'it is *true* that p' is regarded as entailing 'it is certain that p'), this removes the problem for the Rule Perspective. On *this* reading of 'it is certain that p', whenever an agent is entitled to be certain that p (by applying the Principal Principle to the Born Rule probabilities derived from a state she assigns on the basis of the rules governing state assignment) she may well endorse the claim 'I am certain that *p and it is certain* that *p'*. So, the QBMP does not arise.

In response to this line of thought critics may propose a different reading of 'it is certain that *p'*, according to which impersonal certainty is not entailed by truth and according to which the adherent of the Rule Perspective may seem to be committed to the endorsement of QBMP-type sentences after all. One may argue that being *rationally* certain that *p* requires being able to exclude that there are bits of evidence which, if one had access to them, would make one stop being certain that *p*. This requirement may appear to be violated in the Rule Perspective, where an ascription of probability 1 must potentially be revised (and the certainty that may go with it be abandoned) whenever new relevant evidence about which NQMCs are true is obtained. It may therefore seem that agents who accept the Rule Perspective and perform an ascription of Born Rule probability 1 are committed to a statement of the following form (where 'MQBMP' stands for 'modified quantum Bayesian Moore's paradox'):

> **MQBMP:** 'I am certain that *p*, but there might be potential additional evidence which, if I acquired it, would make me stop being certain that *p*.'

Irrespective of how problematic such a statement really is, it would probably not be good news for the Rule Perspective if its adherents were committed to the systematic endorsement of it.

Fortunately, this is not the case. To see why, it is useful, first, to recall that the Rule Perspective regards entropy maximisation as a constitutive rule of state assignment (to be applied whenever a quantum state is assigned to a system for the first time) and, second, to observe that entropy maximisation helps avoid situations where an agent must revise an ascription of probability 1 in the light of additional evidence. As explained by Jaynes, what distinguishes an entropy-maximising probability function from its alternatives is that 'no possibility is ignored; it assigns positive [probability] weight to any situation that is not absolutely excluded by the given information' (Jaynes [1957a], p. 623). Similarly, what distinguishes entropy-maximising quantum states from *non*-entropy-maximising ones is that entropy-maximising ones ascribe probability *strictly less than one* to any NQMC '*p*' which is not entailed by the information about true NQMCs on grounds of which the state assignment is made.

As a consequence, whenever an agent obtains information about some quantum system that requires revision of an ascription of Born Rule probability 1, she will interpret it as pertaining to the system in a different physical situation, where the information about true NQMCs which she used to make her original probability ascription is no longer valid. For example, if an assignment of $| \uparrow_z \rangle$ is made on the basis of knowledge that the NQMC 'The value of S_z is $+1/2$' is true, then, if S_x is measured, the ascription of probability 1 to that NQMC must be revised. The revised probability ascription, however, must be read as pertaining to the system *in a different physical situation* than before. Consequently, whenever an adherent of the Rule Perspective is certain, on grounds of a correctly performed state assignment, that some NQMC 'p' holds true of the system at t, she can also be certain that any additional evidence which, if she had it, would lead her not to be certain that 'p' would concern the system under different conditions than those which she knows to obtain. Thus, an adherent of the Rule Perspective would not endorse any statement of the form (MQBMP) with both occurrences of 'p' pertaining to the same system under the same conditions. To conclude, the quantum Bayesian Moore's paradox does not arise as a problem for the Rule Perspective on any of the paradox's possible readings.

Part III
Objections

7
Copenhagen Reloaded?

As announced in the introductory chapter, one of the core motivations for the therapeutic approach developed here is to investigate whether the doctrine of the Copenhagen interpretation, that quantum theory is fine as it stands and needs neither completion nor revision nor reinterpretation in speculative terms, can be vindicated. Having outlined the essential ingredients of the Rule Perspective in the previous chapters as an account designed to fulfil these therapeutic ambitions, it is natural to ask whether the Rule Perspective may be nothing else than the Copenhagen interpretation in new clothing. Since the Copenhagen interpretation is nowadays widely regarded as a failure, the question amounts to a criticism of the Rule Perspective and, so, deserves to be answered in the part of this work that addresses objections.

The two sections of this chapter comment on the parallels and differences between the remarks of Bohr and Heisenberg, widely considered as the two central proponents of the Copenhagen interpretation, and the core ideas of the Rule Perspective, as laid out in the previous chapters. Bohr is usually regarded as the chief authority concerning what counts as the Copenhagen interpretation. For example, this is how the *Stanford Encyclopedia* article (Faye [2008]) on the Copenhagen interpretation sees him. By contrast, Don Howard (Howard [2004]) argues convincingly that Bohr did not share many of the core ideas usually associated with the Copenhagen interpretation. According to him, the expression 'Copenhagen interpretation' itself was coined and employed by Heisenberg with the aim of pretending the existence of a shared perspective on quantum theory held by him, Bohr, and others, which in fact never did exist. Mara Beller concurs and goes as far as denying that *any* coherent combination of the ideas associated with the Copenhagen interpretation was ever held by anyone.[50]

Even if one may not want to go as far as Beller, it is difficult not to acknowledge that it is very difficult to pin down the Copenhagen interpretation as a well-defined, clear-cut view of quantum theory. Therefore, I cannot compare the Rule Perspective with the Copenhagen interpretation in any canonical version and must rather discuss the views of Bohr and Heisenberg (to which I confine myself here) separately.

However, as an additional difficulty, Bohr's remarks on quantum theory in general and those on quantum states in particular are unfortunately highly ambiguous and unclear. This makes it difficult to interpret them, and the question of in which regards Bohr and the Rule Perspective concur and in which they differ is not easily answered unambiguously. In view of this difficulty, I confine myself to a brief comparison of some of Bohr's and the Rule Perspective's core claims where a comparison seems to suggest itself.

At first sight, Heisenberg's remarks on the interpretation of quantum theory appear to lend themselves more readily to a comparison with the Rule Perspective since he seems to commit himself openly to an epistemic account of quantum states – in some passages at least. However, as my discussion will show, there is more than one possible reading of his claims, and the different readings possible tie in different directions as far as even the rather straightforward question of whether he defends an epistemic account of quantum states is concerned. I shall argue that there is at least one reading of his remarks according to which his view comes out similar in some respects to the Rule Perspective. However, his position according to this reading differs strongly from the collection of ideas usually associated with the Copenhagen interpretation. In conclusion, on all readings of the central remarks by the most important 'Copenhagenians' that seem viable the Rule Perspective is no mere 'Copenhagen reloaded'.

7.1 Bohr and 'classical language'

The most evident parallel between Bohr's perspective on quantum theory and the Rule Perspective concerns the distinctions they make between how quantum states are used (namely, in some sense non-descriptively) as opposed to how (what Bohr refers to as) 'classical concepts' are used, namely, according to him, in a 'pictorial' way (Bohr [1962], p. 2). Bohr never seems to link quantum states in any way to the epistemic conditions of those who assign them, but in virtue of the fact that he consistently refers to quantum states as merely 'symbolic', it does not seem unreasonable to attribute to him what in the terminology used in the present work is a 'non-ontic' account of quantum states. Even though that reading seems plausible, one must be careful with it, for he never really seems to put any emphasis on the point that – as is crucial according to non-ontic accounts of quantum states – there is no such thing as the *true* quantum state of a quantum system.

Differences between Bohr's view and the Rule Perspective do not only arise from a difference in emphasis concerning the role of quantum states, but also from their differing conceptions of the other type of language used in quantum theory – the 'pictorial', classical language according to Bohr and the 'NQMCs' according to the Rule Perspective. As explained in the previous chapter, the Rule Perspective follows Healey in refraining from calling NQMCs 'classical' to prevent the misleading impression that, in Healey's words, a NQMC 'carries with it the full content of classical physics' (Healey [2012a], p. 740). In contrast to this word of caution, Bohr accords to classical physics *itself* an ineliminable role in the application of quantum theory. According to him, even though classical physics never ceases to 'represent ... an idealization' (Bohr [1962], p. 2), its employment cannot be avoided in the description of the experimental arrangement in which the quantum system is placed, which, as he emphasises, may not legitimately be ignored.

Erhard Scheibe refers to Bohr's central idea concerning the ineliminable role of classical physics in the application of quantum theory as the 'buffer postulate', which, according to him, is best viewed as the claim that '[t]he description of the apparatus and of the results of observation, which forms part of the description of a quantum phenomenon, must be expressed *in the concepts of classical physics* (including those of 'everyday life'), eliminating consistently the Planck quantum of action' (Scheibe [1973], p. 24; see the references to Bohr's writings given there). The Rule Perspective concurs with the buffer postulate inasmuch as it contends that measurement results (just as any sentences stating facts about quantum systems) are described in terms of NQMCs rather than quantum states. There is no need, however, to eliminate occurrences of Planck's constant $\hbar$ from the NQMCs, for instance in the cases of NQMCs concerning the values of angular momentum or spin.

More importantly, the Rule Perspective does not subscribe to Bohr's repeatedly forwarded claim that *any* application of quantum theory requires that an – albeit context-dependent[51] – line be drawn between the quantum system on the one hand and the experimental context, treated 'classically', on the other. In his own words, 'the introduction of a *fundamental distinction between the measuring apparatus and the objects under investigation* [is] [...] the essentially new feature in the analysis of quantum phenomena' (Bohr [1962], p. 3). As Bohr sees it, this feature fundamentally sets off quantum theory from classical physics. The Rule Perspective may be characterised as accepting a refined version of this claim in that it conceives of environment-induced decoherence – which typically arises through the interaction of the measured systems with those that we call 'apparatuses' – as enabling the ascription of probabilities to the NQMCs that are 'licensed' for the measured system. It denies, however, that to the contrast between quantum states on the one hand and NQMCs on the other corresponds an analogous contrast between measured system and apparatus. Applying quantum theory without singling out anything as playing the role of 'the apparatus' is perfectly possible and coherent – while there is no way of applying quantum theory that does not involve NQMCs (because then there would be no bearers of the quantum probabilities). Thus, the Rule Perspective has no problems with the application of quantum theory in, say, cosmological or astrophysical contexts where there is no obvious candidate 'apparatus' or 'measurement setup'. This is surely a decisive advantage over Bohr's position.

Landau and Lifshitz, in their seminal textbook on quantum mechanics, concisely point out the awkward role of classical physics according to the Bohrian interpretation, which they nevertheless seem to endorse without reservation. According to them, '[q]uantum theory occupies a very unusual place among physical theories: it contains classical mechanics as a limiting case, yet at the same time it requires this limiting case for its own formulation' (Landau and Lifshitz [1977], p. 3). It seems doubtful to me whether this account of the relation between quantum theory and classical physics is coherent at all, and to refer to the relation, thus conceived, as 'unusual' seems quite euphemistic. The Rule Perspective offers a very different account of that relation in that it does not draw the distinction between quantum states on the one hand and NQMCs on the other as a distinction between *quantum* and *classical* features of language at all. NQMCs, from this point of view, are not somehow 'idealizations', as classical physics is according to Bohr, but descriptions of (the properties of) quantum systems in the most straightforward sense, which, however, as explained before, are not all licensed for the same quantum system in all circumstances. As will be further discussed in more detail in Section 9.1, by adopting this point of view the Rule Perspective consistently avoids any commitment to an unappealing ontological micro/macro divide.

Notoriously, the concept of *complementarity* occupies centre state in Bohr's thoughts about quantum theory. Bohr employs it in various, partly obscure, ways, and its actual meaning and significance are extremely difficult to understand.[52] One of its central applications is to what is widely known as 'wave/particle duality', as spectacularly manifest in interference experiments such as the famous two-slit setup; another refers to the contrast between, as Bohr puts it, 'space-time concepts and the laws of conservation of energy and momentum' (Bohr [1934], p. 11). I fail to see how Bohr's employment of the concept of complementarity may contribute substantially to our understanding of quantum theory and see no connection worth emphasising to any ideas underlying the Rule Perspective.

The following section turns to the comparison between the Rule Perspective and Heisenberg's take on quantum states.

7.2 Heisenberg and the epistemic conception of quantum states

Heisenberg is sometimes named as one of the first – if not *the* first – proponent of the epistemic conception of quantum states.[53] In the present section I discuss those passages which suggest such a reading and then some others that seem to make more sense if one attributes to him a propensity interpretation of quantum probabilities. Here I neglect one further possible reading I have outlined elsewhere[54] for its lack of systematic interest. In that reading, Heisenberg is best conceived of as a philosophical opportunist – at least as far as quantum states are concerned – and does not offer us any coherent and well-defined view of quantum states at all.

7.2.1 Heisenberg on pure versus mixed states

The main reason for ascribing to Heisenberg an epistemic account of quantum states is that he relates measurement collapse to a sudden change in the epistemic situation of the observer, claiming that '[s]ince through the observation our knowledge of the system has changed discontinuously, its mathematical representation also has undergone the discontinuous change and we speak of a "quantum jump"' (Heisenberg [1958], p. 28). Since any perspective on measurement collapse as reflecting a change in the epistemic situation of the agent assigning the state presupposes the epistemic conception of states (in whatever specific version), the quote just given provides evidence that Heisenberg subscribes to that view. The way he introduces the notion of a quantum state ('probability function') to his readers can be taken to confirm that reading:

> The probability function represents a mixture of two different things, partly a fact and partly our knowledge of that fact. It represents a fact insofar as it assigns at the initial time the probability unity (i.e. complete certainty) to the initial situation: the electron moving with the observed velocity at the observed position; 'observed' means observed with the accuracy of the experiment. It represents our knowledge insofar as another observer could perhaps know the position of the electron more accurately. The error in the experiment does – at least to some extent – not represent a property of the electron but a deficiency in our knowledge of the electron. Also this deficiency of knowledge is expressed in the probability function. (Heisenberg [1958], p. 19)

Quantum states, according to this statement, depend on the accuracy of the state-assigning observers' knowledge of the systems states are assigned to, and in that sense they have an important subjective element. It may seem more difficult to say what Heisenberg means by suggesting that quantum states also have an objective element – 'a fact', as he writes – that manifests itself in the assignment of probability 1 to the values of observables that one knows after having measured them. The following quote, taken from the same work, appears to indicate that he endorses an epistemic account only for *mixed* quantum states and regards *pure* states as describing objective properties of the systems they are assigned to:

> The probability function combines objective and subjective elements. It contains statements about possibilities or better tendencies ('potentia' in Aristotelian philosophy), and these statements are completely objective, they do not depend on any observer; and it contains statements about our knowledge of the system, which of course are subjective in so far as they may be different for different observers. In ideal cases the subjective element in the probability function may be practically negligible as compared with the objective one. The physicists then speak of a 'pure case'. (Heisenberg [1958], p. 27)

Since Heisenberg argues that the 'subjective element' of quantum states is absent in what he calls 'pure cases', it may seem as if he defended a version of the ignorance interpretation of mixed states, according to which pure states are objective properties of quantum systems, whereas mixed states are only there to express the incomplete knowledge of agents who are ignorant of actual pure states. However, as argued in Section 3.1.2, the ignorance interpretation of mixed states fails to resolve the measurement problem for rather trivial reasons, in that it conceives only of 'properly' mixed states as expressing ignorance, not of 'improper mixtures', as we encounter them in the analysis of the measurement problem (see Eq. (3.6)).

There is textual evidence, however, that Heisenberg intends his remarks on mixed states as expressing some sort of 'ignorance' indeed as applying to improper mixtures after all, since he emphasises the role of the coupling between the measured system and the measuring apparatus, claiming that 'it is very important to realize that our object [being measured] has to be in contact with the other part of the world, namely, the experimental arrangement, the measuring rod, etc., before or at least in the moment of observation' (Heisenberg [1958], p. 27). He clearly seems to ascribe a 'subjective element of incomplete knowledge' to the (improperly) mixed state assigned to the measured system when he writes that '[a]fter this [measurement] interaction has taken place, the probability function contains the objective element of tendency and the subjective element of incomplete knowledge, even if it has been a "pure case" before' (Heisenberg [1958], p. 28). This 'incomplete knowledge', however, as we have seen, cannot be incomplete knowledge about any alleged *true*, pure state the measured system is in, so, we might ask, what kind of incomplete knowledge can it be? Heisenberg's answer to this question is that it is incomplete knowledge about the microscopic details of the rest of the world to which the measured system, via the measurement device, must be coupled. Thus, the 'subjective element' which, according to himself, is present in (improperly) mixed states seems to relate to an ignorance on behalf of the observer in that it reflects her 'uncertainties of the microscopic structure of the whole world' (Heisenberg [1958], p. 27).

It is unclear, however, why interpreting improper mixtures as reflecting ignorance of 'uncertainties of the microscopic structure of the whole world' might help us justify collapse as envisaged. A measurement supposedly provides information about macroscopic features of the apparatus as well as of the measured system, not about the 'microscopic structure' of the rest of the world. Even if it did, it would remain unclear why this information might require the change in the state assigned to the system being measured in the way prescribed by von Neumann's projection postulate. I confess that I am unable to make sense of these remarks of his, which, on the face of it, appear to be highly relevant for the adequate understanding of his interpretation of quantum states.

7.2.2 Heisenberg on quantum probabilities as 'objective tendencies'

Independent of what the correct interpretation of Heisenberg's remarks on the subjective element in mixed states really is, his statement that the quantum state 'contains statements about possibilities or better tendencies ("potentia" in Aristotelian philosophy) [which] are completely objective [in that] they do not depend on any observer' presents formidable problems for any reading of him as a proponent of the epistemic conception of quantum states. For what he seems to endorse in this passage is an account of quantum probabilities as objective along the lines of propensity interpretations, as outlined in Section 6.2.2.

Henry P. Stapp proposes an interpretation of Heisenberg which explicitly adopts such a reading by neglecting Heisenberg's remarks on collapse as reflecting a change in the observer's epistemic situation, and conceiving of it as a physical process that from time to time interrupts unitary time-evolution as governed by the Schrödinger equation in an irregular manner. According to this interpretation of Heisenberg, '[t]he fundamental dynamical process of nature is no longer one single uniform process, as it is in classical physics. It consists rather of two different processes' (Stapp [2003], p. 41); that is, the smooth, unitary time-evolution and the discontinuous state-change of collapse are both regarded as physical processes. The features of reality that correspond to the sudden state-changes of collapse are referred to by Stapp as 'actual events'. Quantum mechanical probabilities are objective propensities in this view in that they measure the object-inherent tendencies of the actual events to occur. The picture of reality that emerges from this interpretation exhibits similarities to GRW theory, as reviewed in Section 3.3.2 (in particular in the 'flash version'), which complements time-evolution according to the Schrödinger equation with randomly interspersed spontaneous localisation processes. In comparison to GRW theory, however, the decisive disadvantage of Heisenberg's position as construed by Stapp is that it does not give us a formal and quantitative account of the frequency and dynamics of the 'actual events', even though they are regarded as the physical happenings *par excellence*.

Stapp's interpretation of Heisenberg's remarks on quantum probabilities is not only unsatisfying for systematic reasons but also on exegetical grounds. In particular, it identifies 'the transition from the "possible" to the "actual"' (Heisenberg [1958], p. 28), repeatedly emphasised by

Heisenberg, with the discontinuous change of the quantum state in measurement collapse, even though Heisenberg categorically claims that it has nothing to do with that change, which in turn results from the 'discontinuous change of our knowledge in the instant of registration' (Heisenberg [1958], p. 29).

Kristian Camilleri, in his recently proposed novel interpretation of Heisenberg's later writings, argues that the transition 'from the "possible" to the "actual"' must be conceived of as one between two different forms of description, not as one between two different types of metaphysical categories (as suggested by Stapp). In addition, as Camilleri convincingly argues, this transition is not connected with measurement collapse:

> The 'transition from the potential to the actual' is therefore completely misunderstood if it is interpreted as a collapse of a physically real extended wave-packet in space. Nor should it be interpreted in Berkeley's sense: *esse est percipi*. Rather, we must understand the 'actual' and the 'possible' as two modes of description, both of which employ the language of time and space at some level. The transition from potentiality (a quantum-mechanical description) to actuality (a classical space-time description) must be understood as a transition from one *mode of description* to another. The two modes of description – the possible and the actual – are deemed complementary in Heisenberg's, though not in Bohr's, sense of the term. (Camilleri [2009], p. 170)

Conceived of as a contrast between two different types of language, the distinction between 'actuality' and 'potentiality' becomes closer to the distinction between NQMCs and quantum states, which takes centre stage in the Rule Perspective. As between Bohr's account and the Rule Perspective, however, there remains an important difference between Heisenberg's account as conceived of by Camilleri and the Rule Perspective in that, according to Heisenberg more or less as according to Bohr, 'a dividing line must be drawn between, on the one hand, the apparatus which we use as an aid in putting the question and thus, in a way, treat as part of ourselves, and on the other hand, the physical systems we wish to investigate [which] we represent mathematically as a wave function' (Heisenberg [1979], p. 15). In other words, the system/apparatus divide, conceived of as directly corresponding to the distinction between the two forms of description, is seen as essential for the application of quantum theory by Heisenberg, while it is not regarded as important by the Rule Perspective.

As these remarks show, Heisenberg's remarks on the nature of quantum states are rather elusive, and it is therefore difficult to decide whether his position is close to the Rule Perspective or not. One diagnosis of the reason why his perspective on quantum states is so difficult to pin down is that there is just no clear-cut position behind his remarks and his apparent core claims about quantum states are made mainly for rhetorical purposes. I suggest a reading of Heisenberg's remarks on quantum states along these lines in (Friederich [2013a], section 6) without claiming that it has to be the correct one, but I will not repeat it here, since the present, more systematically oriented discussion would not substantially profit from it. The next chapter turns to criticisms of the Rule Perspective that go beyond the charge that it is 'Copenhagen reloaded'.

8
The Charge of Anthropocentrism

8.1 Bell's criticism

Epistemic accounts of quantum states such as quantum Bayesianism and the Rule Perspective are formulated in terms of notions such as 'agent', 'epistemic situation', and 'state assignment', which are neither themselves fundamental physical notions nor in any evident way reducible to such notions. In the present section I discuss an objection to the employment of such *'anthropocentric* notions' (as I will call them), which claims that they are far too vague and too imprecise to be used in foundational accounts.

First, let us take a closer look at the role of anthropocentric notions in epistemic accounts of quantum states. The notion of an *agent*, to begin with, is employed in the statement of the epistemic conception of quantum states itself, claiming that quantum states reflect the assigning agents' epistemic conditions with respect to the system states are assigned to. Clearly, this idea cannot be expressed without using some notion of a subject who has an epistemic condition and assigns a quantum state. Furthermore, 'epistemic condition' and 'state assignment' are themselves no less anthropocentric than 'agent' (or 'subject'). Both appear equally irreducibly and ineliminably in the statement of the epistemic conception of states itself. In addition, some relative of the notion of measurement is employed in both quantum Bayesianism and the Rule Perspective: in the case of quantum Bayesianism the relevant notion is that of an 'experimental intervention [...] into nature' (Fuchs [2002], p. 7); in the case of the Rule Perspective it is that of an event resulting in 'new knowledge' of the values of observables, to be taken into account when updating one's state assignment in accordance with Lüders' Rule. Different epistemic accounts of quantum states may use different anthropocentric notions, but in view of the crucial role which anthropocentric notions play in the statement of the epistemic conception of states itself and in the more detailed considerations of the two versions discussed here it seems highly plausible that they cannot dispense with them altogether.

Interpretations of quantum theory that rely on anthropocentric notions are heavily criticised by some of the most distinguished interpreters of the theory. J. S. Bell, for instance, claims that such notions are not sufficiently sharp and fundamental to be used in foundational accounts. Among the words which, according to him, 'however legitimate and necessary in application, have no place in a *formulation* with any pretension to physical precision' (Bell [2004], p. 209; the emphasis is Bell's), one finds, for instance, 'observable, information, measurement', which, evidently, are close relatives of the anthropocentric notions encountered in the epistemic accounts of quantum states discussed here. Bell's main complaint concerns the role of these notions in accounts which postulate the occurrence of measurement collapse whenever a quantum system is measured. As already quoted in Section 2.2, Bell comments sarcastically on this idea:

> What exactly qualifies some physical systems to play the role of 'measurer'? Was the wavefunction of the world waiting to jump for thousands of millions of years until a single-celled living creature appeared? Or did it have to wait a little longer, for some better qualified system […] with a PhD? If the theory is to apply to anything but highly idealised laboratory operations, are we not obliged to admit that more or less 'measurement-like' processes are going on more or less all the time, more or less everywhere? Do we not have jumping then all the time? (Bell [2007], p. 34)

Bell's criticism of the textbook account of measurement collapse as occurring whenever the system is measured is certainly reasonable and adequate. However, it does not directly apply to the versions of the epistemic conception of quantum states considered here. These deny that there is such a thing as *the* wave function of a quantum system with respect to which the question of when it 'jumps' could meaningfully be asked. In particular, according to these accounts there exists no such thing as a 'wave function of the universe', which at one point 'jumps' for the first time in its history. Primitive anthropocentric notions are indeed used, but their role is not that of specifying under which conditions which physical process takes place but to characterise the different conceptual roles of the elements of the quantum theoretical formalism, in particular whether they are used to represent features of reality or not.

The difference between these two different kinds of appeal to anthropocentric notions is clarified by Fuchs in his quantum Bayesian characterisation of quantum theory as a 'users' manual that *any* agent can pick up and use to help make wiser decisions in this world of inherent uncertainty' (Fuchs [2010], p. 8). In this picture of quantum theory, the theory is construed as an empirical extension of subjective Bayesian probability theory, where probability theory, in turn, is construed as an 'extension of logic' (Fuchs [2010], p. 8, fn. 14). Fuchs argues convincingly that based on this conception quantum Bayesianism cannot be plausibly asked to deliver a reductive analysis of notions such as 'agent' any more than philosophers of logic can be asked to deliver a reductive account of a notion such as 'logical subject' (that is, the notion of an agent who employs the methods of logic):

> [I]s [...] quantum mechanics [...] obligated to derive the notion of agent for whose aid the theory was built in the first place? The answer comes from turning the tables: Thinking of probability theory in the personalist Bayesian way, as an extension of formal logic, would one ever imagine that the notion of an agent, the user of the theory, could be derived out of its conceptual apparatus? Clearly not. How could you possibly get flesh and bones out of a calculus for making wise decisions? The logician and the logic he uses are two different substances – they live in conceptual categories worlds apart. One is in the stuff of the physical world, and one is somewhere nearer to Plato's heaven of ideal forms. Look as one might in a probability textbook for the ingredients to reconstruct the reader himself, one will never find them. So too, the Quantum Bayesian says of quantum theory. (Fuchs [2010], pp. 8f.)

As becomes clear in this passage, according to Fuchs' quantum Bayesian take on quantum theory the theory has an essentially normative role in that it functions as a 'manual' to 'make wiser decisions'. That manual is formulated in terms of notions such as that of a quantum state and, using these notions, recommends certain ways of forming one's credences about outcomes if one wants to get along well in our world. Now, if quantum theory's essential role is indeed that of a manual that guides our credences and behaviour, one can hardly expect an account of what constitutes a quantum theoretical agent that is phrased solely in terms of the notions that are used in the manual itself. This does not mean that one cannot characterise the notion of an agent availing herself of the methods laid down in that manual in terms of *other*, richer, bits of language. But it should not come as a surprise that these bits of language will have to involve anthropocentric notions and not just those of fundamental physics.

Fuchs' comparison between the roles of the agent in quantum theory and logic might be criticised on the grounds that quantum theory, unlike logic, is a physical theory and that agents using quantum theory and the world they live in are themselves (aggregates of) objects studied in physics, whereas agents availing themselves of the methods of logic and the world they live in are not (aggregates of) objects studied in logic (whatever exactly one takes these to be). The crucial point of Fuchs' argument, however, is that in quantum Bayesianism anthropocentric notions are employed at the *meta-level* of characterising the status of quantum theoretical concepts – claiming that quantum states are used non-descriptively, for instance – not at the *object-level* of describing physical processes themselves – such as physical collapse and under which conditions it occurs. The textbook accounts criticised by Bell, by contrast, make object-level use of anthropocentric notions in that they conceive of measurement collapse as a physical process that takes place whenever the system is measured.

To sum up, Bell's verdict against anthropocentric notions in foundational accounts seems reasonable only with respect to accounts which invoke these notions at the object-level of physical processes. There is no reason to extend it to accounts which use anthropocentric notions to clarify how the different elements of the quantum theoretical formalism connect to physical reality if the formalism is correctly applied. Inasmuch as this is what quantum Bayesianism and the Rule Perspective do, they are not concerned by Bell's criticism. The following subsection investigates whether epistemic accounts of quantum states can confine their appeal to anthropocentric notions to the meta-level of conceptual clarification even when they try to answer the challenge of why those observables which we 'measure' have always determinate values.

8.2 Anthropocentric notions and value determinateness

In debates about the foundations of quantum theory it is often claimed that appeal to 'measurement' should not be understood as entailing the existence of a value of the quantity taken to be 'measured' prior to the measurement act. The idea that a quantum measurement is an act of *creation* rather than an act of establishing what is there has a long tradition. Pascual Jordan, for instance, endorses it when he writes that it is 'we ourselves [who] bring about the matters of fact' (Jordan [1934], p. 228, my translation) which we usually think of as being determined in experiments. Similar ideas can be found in the writings of adherents of the epistemic conception of quantum states, for instance in those of the quantum Bayesians Fuchs and Schack, who claim that the 'measured values' of observables are not merely '"read off"' in measurement but rather 'enact[ed] or creat[ed] [...] by the [measurement] process itself' (Fuchs and Schack [2009], p. 3).

Quantum Bayesianism's main motivation for denying the existence of determinate values prior to measurement is the conflict between assuming such values and the famous no-go theorems on determinate value assignments originating from Gleason, Bell, Kochen, and Specker.[55] As explained in Section 5.1, quantum Bayesianism denies that we can ever know which observable some numerical value obtained in an experiment is a value of, but it does not go as far as claiming that observables do not have any values at all. As a result, quantum Bayesianism is drawn to the conclusion that determinate values are created in the act of measurement. In the words of Fuchs:

> QBism [i.e., quantum Bayesianism as defended by Fuchs] says when an agent reaches out and touches a quantum system – when he performs a quantum measurement – that process gives rise to birth in a nearly literal sense. With the action of the agent upon the system, the no-go theorems of Bell and Kochen–Specker assert that something new comes into the world that wasn't there previously: It is the 'outcome,' the unpredictable consequence for the very agent who took the action. (Fuchs [2010], p. 8)

And a few pages later:

> That the world should violate Bell's theorem remains, even for QBism, the deepest statement ever drawn from quantum theory. It says that quantum measurements are moments of creation. (Fuchs [2010], p. 14)[56]

The problem with any conception of 'quantum measurement' as a type of 'birth' or 'creation' is that it seems to lead straightforwardly to an application of 'measurement' in what according to the terminology used in the previous subsection counts as an *object-level* type of way. In this case, it is not the occurrence of collapse as a physical process which is taken to occur when a measurement is performed, but the 'birth' of the value of an observable when it is measured. According to these considerations, it seems that quantum Bayesianism is unable to confine its employment of anthropocentric notions to the meta-level of conceptual clarification alone, in which case Bell's criticism applies after all.

From a general perspective, the challenge of accounting for determinate post-measurement values without relying on object-level anthropocentric notions can be construed as a re-appearance of the measurement problem in disguise: as explained in Chapter 4, the original measurement problem is dissolved in epistemic (or, more generally, non-ontic) accounts of quantum states by rejecting the notion of the quantum state a quantum system is in, but the question as to why measurement processes, as a matter of fact, do always result in determinate outcomes remains unaddressed. It is tempting to read the passages of the quantum Bayesians on 'quantum measurement' as a type of 'birth' or 'creation' as a response to this challenge that invokes primitive anthropocentric notions at the object-level. In that case, however, Bell's criticism applies to it and its dissolution of the measurement problem cannot be considered successful.

In defence against this charge, quantum Bayesians may point out that to conceive of measurement as an act of 'birth' of the value of an observable is not the same as using 'measurement' to *define* under which conditions determinate values are assumed. According to Fuchs, 'for the QBist, the real world [...] – with its objects and events – is taken for granted' (Fuchs [2010], p. 7). Quantum Bayesianism, one may take this to be saying, *presupposes* the existence of 'events' where determinate values are assumed as a primitive fact, which is not in need of any further grounding. They might add that among these events are the 'measurement events' – namely, those which we use to obtain information as to what the values of observables really are. Phrased differently (and perhaps somewhat crudely): observables do not assume determinate values *because* some process is a measurement process (an idea analogous to the one ridiculed by Bell), but we *call* certain processes measurement processes because the values of observables taken to be 'measured' are determinate at their end (in such a way that we can obtain information about them). Following this line of thought, the quantum Bayesian account of 'measurement' as an act of 'birth' does not necessarily lead to the employment of anthropocentric notions at the object-level.

What quantum Bayesianism does not account for, however, is how physicists can be *confident* as regards under which conditions which observables have determinate values. The Rule Perspective is in better shape than quantum Bayesianism here: it says that competent users of quantum theory are entitled to apply the Born Rule to the values of those observables with respect to which the density matrix they assign is (at least approximately) diagonal in a preferred basis determined by the decoherence effects that are taken into account. This means that these agents, in effect, are entitled to presuppose the existence of sharp values corresponding to those that are ascribed nonzero Born Rule probabilities for precisely these observables.

To see this, consider a measurement setup that is modelled, as before, as a two-component system by consisting of a 'measured system' on the one hand and an 'apparatus' on the other. What it takes for a two-component system to qualify as a candidate 'measurement setup', apart from the fact that the apparatus must be macroscopic so that the result can be read off the display, is that effects from environmental decoherence, when they are taken into account, will render the density matrix assigned to the combined system approximately diagonal in an eigenbasis of $S \otimes A$, where S is the 'measured observable' and A the 'pointer observable', which distinguishes between the macroscopically distinct apparatus states. As the Rule Perspective claims (following Healey), quantum theory in this case 'licenses' those NQMCs which claim that the values of S and A lie in ranges Δ_S and Δ_A of possible values of S and A. From the point of view of the Rule Perspective, to treat a NQMC of the form 'The value of A lies in Δ_A' as 'licensed' by quantum theory means to proceed by assuming that the value of A is either determinately inside Δ_A or outside Δ_A. Quantum theory thus gives the agent entitlement to treat the values of both the 'measured' and the 'pointer' observables as determinate (within bounds as determined by decoherence), so operating under the assumption that these observables do *not* have determinate values (whatever this would practically mean) would mean *failure* to apply the theory correctly.

The Rule Perspective thus not only dissolves the measurement problem by undermining its formulation in terms of quantum states quantum systems 'are in' but goes on to claim that determinateness of the values of the 'measured' observable is to be presupposed by competent users of the theory. If an experimental setup, designed to measure the value of S, is such that taking into account decoherence effects does *not* provide entitlement to apply the Born Rule to those NQMCs that attribute sharp values to S, the setup simply does not qualify as a candidate 'measurement setup' for S. In other words, verifying that the setup qualifies as a 'measurement setup' for S suffices for ascertaining that one has an entitlement to assume that the value of *S* is determinate in it. Without employing any primitive anthropocentric notions at the object-level this accounts for why outcome determinateness is something that physicists must assume if they are to apply quantum theory correctly.

These remarks are neutral with respect to the question of whether even those observables have determinate values for which the NQMCs making claims about their values are *not* licensed for any (actual or hypothetical) agent assigning a quantum state to the system. The application of quantum theory in practice does not itself dictate any specific answer to this question; so, in accordance with the therapeutic ambitions underlying the Rule Perspective, I adopt a minimalist reading here which neither claims that such values exist nor that they do not exist. The prospects for assuming that sharp values exist for all observables – not just for those for which the associated NQMCs are 'licensed' – are investigated in Chapter 12.

9
Reduction and Explanation

9.1 The micro/macro divide

One of the defining aspects of the epistemic conception of quantum states is that it conceives of quantum states as non-descriptive. In this section, I consider the criticism that *any* non-descriptive reading of quantum states entails an unattractive ontological quantum/classical divide. This divide is supposed to separate a realm of classical macro-objects, which are susceptible to descriptions in terms of classical concepts, from a realm of quantum micro-objects, for which no descriptive account can be given at all.

Timpson makes the claim that such an ontological divide is a natural consequence of quantum Bayesianism, but does not regard this as a devastating objection to that position. To substantiate the claim that the idea of an ontological quantum/classical divide is coherent, he proposes an ontological framework which comprises, on the one hand, a 'micro-level we have dubbed unspeakable to which we are denied direct descriptive access' (Timpson [2008], p. 597) and, on the other hand, a 'macroscopic or classical level [which] will be a level of objects which do have unproblematically stateable truths about them' (Timpson [2008], p. 598).

Any such two-level ontological framework is confronted with obvious difficulties: is the idea that reality is a patchwork which consists of a describable and an indescribable ('unspeakable') ontological level really intelligible? And, if it is, where exactly should the line be drawn between the two levels? Drawing on ideas from Nancy Cartwright's philosophy of science, Timpson suggests that the framework can indeed be made coherent if one admits at the macro-level 'metaphysically emergent properties [which] a composite can possess but which its components cannot[,] and which are not conferred on it by the properties possessed by its components and the laws (if any) which they obey' (Timpson [2008], p. 599). As he seems to suggest, accepting such properties and their anti-reductionist ramifications is the price to be paid for a non-descriptive reading of quantum states. If the quantum Bayesian is willing to pay this price, the challenge can be met: '[i]f called upon, ... the quantum Bayesian seems able to present an intelligible ontology to underlie their position' (Timpson [2008], p. 600).

Others are more sceptical as regards the intelligibility of a two-level ontology comprising a level of objects that admit a descriptive account and a level of objects which don't. Marchildon, for instance, regards the empirically manifest fact that macroscopic objects 'are always in definite states' (Marchildon [2004], p. 1462) as inexplicable on an epistemic account of quantum states. According to him, these accounts have to choose between three equally unattractive claims, namely, that either microscopic quantum objects 'do not exist', that they 'may exist, but have no states', or that they 'may exist and may have states, but attempts at narrowing down their existence or specifying their states are useless, confusing, or methodologically inappropriate' (Marchildon [2004], p. 1462). As he sees it, all these options are implausible, given that the 'unspeakable' (Timpson's expression) micro-objects are the (mereological) constituents of the macro-objects for which unproblematic 'definite states' do exist. Whether or not one regards this criticism as ultimately compelling, the challenge of combining a macro-level of describable 'classical' objects with a micro-level of 'unspeakable' quantum objects in a coherent metaphysical framework remains formidable. However, as I shall argue now, the example of the Rule Perspective shows that a two-level ontology of this kind need not be embraced by epistemic accounts of quantum states.

The Rule Perspective locates the chief difference between NQMCs and quantum states in how they are used and applied to reality (the NQMCs descriptively; the quantum states non-descriptively), not in a contrast between distinct realms of objects they are about. On the one hand, quantum states can legitimately (and with much theoretical gain) be assigned to objects of arbitrarily large 'macroscopic' size, for instance by the help of the many-particle methods of quantum statistical mechanics. Their assignment is not confined to the objects of any putative 'micro-level'. On the other hand, even the most 'microscopic' quantum objects elementary particle physicists have discovered (quarks, leptons, gauge bosons, etc.) can be described very well in terms of NQMCs – with the caveat that not all NQMCs are licensed in all circumstances – just as much as the largest objects of cosmology. There is even an interplay between the application of NQMCs and the assignment of quantum states to one and the same quantum system: agents assign quantum states on the basis of their epistemic relations to the systems – on the basis of what NQMCs they know to be true –, and the 'licensing' of further NQMCs depends on the features of the quantum states they assign by taking into account the effects of environment-induced decoherence. The idea of a system to which *only* quantum states but no NQMCs can be applied makes no sense on this view. The distinction between quantum states and non-quantum claims corresponds to a difference in mode of use, not to a difference between different types of objects referred to.

To conclude, epistemic accounts of quantum states are not committed to an ontological micro/macro divide. The Rule Perspective, in particular, does in no way suggest any two-level ontology where a 'classical' macro-level contrasts with an 'unspeakable' quantum micro-level. Furthermore, as I shall argue in what follows, there is no particular problem for non-ontic accounts of quantum states in accounting for reductive explanation of macro-properties in terms of the behaviour of micro-objects.

9.2 Explanation without ontic quantum states

Quantum theory is perhaps the theory with the greatest explanatory success in all the history of science. The final objection to epistemic accounts of quantum states to be discussed here is that non-descriptive readings of quantum states (that do not come together with a theory about hidden variables) are *in general* unable to account for quantum theory's incontestable explanatory force. Timpson raises this charge against quantum Bayesianism by (rhetorically) asking 'if quantum mechanics is not to be construed as a theory which involves ascribing properties to micro-objects along with laws describing how they behave, can we account for [its] explanatory strength?' (Timpson [2008], p. 600) According to Timpson, the quantum Bayesian interpretation of quantum states as 'states of belief' entails that all that can possibly be accounted for by means of quantum theoretical reasoning involving quantum states are agents' beliefs and expectations, not physical phenomena themselves. As an example, Timpson considers the explanation provided by quantum many-particle theory as to why some solid materials (sodium, in his example) conduct electricity well, whereas others don't, and remarks that '[u]ltimately we are just not interested in agents' expectation that matter structured like sodium would conduct; we are interested in *why in fact it does so'* (Timpson [2008], p. 600). Quantum theory, as Timpson emphasises, helps us explain the behaviour of physical systems only because its vocabulary refers to these systems themselves, not to the scientists and their expectations and beliefs about them. Any interpretation of quantum theory that attempts to account for its explanatory force must respect this.

There is a natural response by means of which quantum Bayesians may attempt to address this challenge. It starts with the observation that on the quantum Bayesian account of the quantum theoretical formalism as a 'manual [...] to help make wiser decisions' (Fuchs [2010], p. 8) the theory does not say anything about what agents actually *do* believe, but rather about what they *should* believe in which circumstances to maximise their predictive success. On this reading, quantum theory does not so much *describe* our expectations and beliefs as *prescribe* what we should expect and believe under which conditions.[57] Quantum Bayesians may therefore suggest that any quantum theoretical reasoning which leads to the expectation that some physical phenomenon or regularity will occur is a candidate quantum theoretical explanation of precisely that phenomenon or regularity, contrary to what Timpson suggests.

This suggestion leads directly into the troubled waters of the complicated debate in the philosophy of science concerning the relation between explanation and prediction, which is too involved to be discussed here thoroughly. To keep matters simple, however, we can proceed by assuming that the quantum Bayesian may simply regard it as a *ramification* of her view that to account for why some phenomenon or regularity is to be expected when correctly applying quantum theory (and in that sense predicting it quantum theoretically) is simply what it *means* to explain that phenomenon or regularity quantum theoretically. In this vein, Richard Healey bases his pragmatist account of 'how quantum theory helps us explain' (the title of (Healey [forthcoming])) on the idea that '[t]o use a theory to explain a regularity involves showing the regularity is just what one *should expect* in the circumstances, if one accepts that theory' (Healey [forthcoming], p. 6, emphasis mine). To apply this idea to the example proposed by Timpson, consider an agent who competently applies quantum many-particle theory and arrives at the expectation that matter having the structure and composition of sodium should conduct electricity well. The quantum theoretical reasoning used is not *about* the agent's expectations and beliefs, but it certainly functions as a guide for the agent when forming them. Why not count it as a candidate quantum theoretical explanation of the phenomenon or regularity at issue? It seems that quantum Bayesians may well claim that they are able to account for quantum theory's explanatory force since they do not construe the theory as describing what agents actually *do* believe and expect, but, instead, what they *should* believe and expect.

Unfortunately, however, this response is not open to quantum Bayesianism, and the main reason why is again the position's rejection of the notion of a state assignment being performed correctly. On the quantum Bayesians' view, it does not make any sense to ask whether some line of reasoning that involves the assignment of quantum states is correct or incorrect, and this means that quantum theory does not actually have the prescriptive bite which the line of defence just considered attributes to it. In other words, quantum Bayesianism does not have the conceptual resources to distinguish between failed and successful explanations in quantum theory. Arguably, this means that it cannot account for quantum theoretical explanation in general, and this vindicates Timpson's charge.

The Rule Perspective has much better resources to account for the notion of a quantum theoretical explanation, as it acknowledges from the start that the question of correctness does apply to lines of quantum theoretical reasoning which may lead to expectations about physical phenomena and regularities. In accordance with Healey's account of explanation in quantum theory, as formulated in (Healey [forthcoming]), it emphasises the importance of NQMCs in describing both the phenomena and regularities for which quantum theoretical explanations are sought (the explananda) and the assumptions and background conditions on which the suggested explanations are based (the explanantia). Regarding the example of explaining the experimentally determined conductivities of solid materials, the explanantia include descriptions of the internal composition and structure of the materials together with information about the external conditions the materials are exposed to. Given these 'known facts' about the systems at issue, the rules of state assignment dictate which quantum states to assign (though approximations may have to be made in practice due to the enormous difficulties involved in computing these states exactly). The Born Rule then permits deriving the probabilities for the possible values of observables from these states or, what amounts to the same, it permits computing expectation values for these observables.[58] One such expectation value is the quantum theoretically predicted conductivity. If it is in agreement with what is measured experimentally, the latter may count as explained by quantum theory – or so proponents of the Rule Perspective might suggest.

However, the idea that a quantum theoretical explanation is what makes an explanandum phenomenon or regularity *expected* can be nothing more than a first approximation to a more accurate picture. In the remaining paragraphs of this section I discuss two respects in which it needs refinement and argue that the Rule Perspective has the conceptual resources to refine it as necessary. The two aspects correspond to two criticisms of Hempel's IS- (*inductive statistical*) model of probabilistic explanation made by Peter Railton (Railton [1978], pp. 211f.) and David Lewis. According to Lewis, Hempel's model has the following two 'unwelcome consequences':

> (1) An improbable event cannot be explained at all. (2) One requirement for a correct explanation [...] is relative to our state of knowledge; so that or ignorance can make correct an explanation that would be incorrect if we knew more. Surely what's true is rather that ignorance can make an explanation seem to be correct when really it is not. (Lewis [1986b], p. 232)

The claim that a quantum theoretical explanation is any line of quantum theoretical reasoning that makes an explanandum phenomenon or regularity expected has the same 'unwelcome consequences' as Hempel's IS-model.

Let us briefly address the second 'unwelcome consequence' first. It seems plausible that a line of reasoning which makes some phenomenon or regularity expected, but which allows for the possibility that this may change once more is known about the system at issue, does not qualify as a candidate quantum theoretical explanation. The Rule Perspective does not have any problems formulating this requirement: the rules of state assignment prescribe what quantum state assignment should be made under which epistemic conditions, and they apply whether or not these conditions are actually realised. In particular, one can meaningfully ask what expectations an agent *should have* if she had whatever *additional* information that is not excluded by the current available data. So, the Rule Perspective can require that a successful explanation need not only make the explanandum expected but that it should, moreover, ensure that the explanandum remain expected if the agent's epistemic situation were to improve in whatever manner conceivable.

Let us now turn to the first 'unwelcome consequence' brought up by Lewis and Railton. It concerns the explainability of improbable events such as decays of radioactive nuclei with long half-lives. For these nuclei, considered individually, their decay is not to be expected on the time scale of a typical human life. There are quantum theoretical computations of the probabilities of these decays, which start from assumptions concerning the structure and interactions between the constituents of these nuclei. One may feel that these computations are explanatorily relevant as regards even the most rare nuclear decays and conclude that this disqualifies any perspective on quantum theoretical explanation that requires that a high probability be conferred on the explanandum.[59]

To respond to this challenge, the proponent of the Rule Perspective may deny that arbitrary single events such as individual radioactive decays are candidate recipients of quantum theoretical explanations at all. Probabilistic theories such as quantum theory, as argued by (Woodward [1989]) along general lines, can in general explain only *patterns* or *regularities* among events. As Healey notes, 'the phenomena that physicists are primarily concerned to explain are not particular individual happenings but general regularities in the properties or behaviour of physical systems of a certain kind',[60] which is why he confines his account of 'how quantum theory helps us explain' to the explanation of regularities.

This perspective still allows us to regard individual events, no matter how low their (nonzero) probabilities, as in an interesting sense *accounted for* by quantum theory, though not as genuinely *explained*. In particular, this concerns individual radioactive decays. Classically, these are forbidden due to a barrier of potential energy which prevents the particles inside the nucleus from leaving it. Quantum mechanically, in contrast, they are allowed due to the phenomenon known as 'quantum tunnelling': the penetration of a system (for example, an alpha particle) through a potential barrier which is larger than the system's total energy, which is just what happens in radioactive decay. Thus, unlike classical mechanics, quantum theory accounts for radioactive decay by predicting that it happens and how often. However, what it genuinely *explains* are the observed decay rates, not the individual decays.

Returning to the Lewis/Railton challenge, however, it seems that we have to admit that here there are some individual events for which quantum theory seems to deliver single-case explanations, and the Rule Perspective should not fail to classify these explanations as successful. To begin with an example, in addition to the *unsolicited* radioactive decays just considered there are also the *stimulated* radioactive decays. In these cases, radioactive decay is *induced* by absorption of a neutron by a nucleus, triggering its decay.[61] It seems natural to regard quantum theory as accounting for how stimulated decays are *caused* by incoming neutrons and as thereby delivering *causal explanations* of these decays. But is there a place for causal explanations using quantum states if quantum states are conceived of along the lines of the epistemic conception?

Arguably, this depends on one's favoured account of causation and causal explanation in particular. According to the Rule Perspective, quantum states apply to quantum systems only in favour of their being assigned by some (actual or hypothetical) agent, characterised by her distinctive epistemic perspective on the quantum systems. Thus, what we need is an account of causal relations as relational with respect to agent perspectives in an analogous way. Fortunately, such accounts of causation exist, most prominently perhaps the *agency theory* championed by Peter Menzies and Huw Price. The agency theory spells out causation in terms of 'agent probabilities', where, as explained by Menzies and Price, 'the agent probability that one should ascribe to B conditional on A (which we symbolize as "$P_A(B)$") is the probability that B would hold were one to choose to realize A' (Menzies and Price [1993], p. 190). The core underlying idea is that A causes B if and only if $P_A(B) > P(B)$ for the 'agent probability' $P_A(B)$. As Price makes clear in a more recent statement of his position, one important aspect of the 'agentiveness' of agent probabilities is 'that they are assessed from the *agent's* distinctive epistemic perspective' (Price [2012], p. 494. The emphasis is Price's). This strongly suggests that quantum probabilities, as conceived of from the point of view of the Rule Perspective, can play the role of agent probabilities in the sense of the agency theory of causation.

Agents applying quantum theory can realise that they may (at least in principle; sometimes as a matter of fact) bring about the conditions in which stimulated radioactive decay occurs. Thus, they may legitimately regard the probability they ascribe on the basis of their quantum theoretical computations to the stimulated decay of a radioactive nucleus as an 'agent probability' in the sense of Menzies and Price. Even if that probability remains low – yet enhanced as compared to a situation where no stimulation via neutron absorption occurs –, the decay counts as 'caused' by the incoming neutron on the agency theory as combined with the Rule Perspective's view of quantum probabilities. We have therefore found a robust sense in which even individual events such as stimulated decays are successfully *explained* through quantum theory by the standards of the Rule Perspective.

'Explanation' is a complicated and multifaceted notion. Therefore, we should not expect to receive an account of what constitutes a quantum theoretical explanation that is fully condensed in a single criterion.[62] However, as the considerations just presented suggest, the Rule Perspective does not have any principled difficulties accounting for the force and genuineness of quantum theoretical explanations. Furthermore, as argued in the previous section, it is not committed to an ontological micro/macro divide and is therefore no less well compatible with reductionist accounts of quantum theoretical explanation than the most-discussed ontic accounts of quantum states. As this shows, epistemic account of quantum states need not go hand in hand with the explanatory anti-reductionism advocated by the quantum Bayesians.[63] Reductive explanation can have a place in non-ontic as much as in ontic accounts of quantum states and is not in any conflict with the therapeutic approach put to work here.

Part IV

Non-locality, Quantum Field Theory, and Reality

10
Non-locality Reconsidered

10.1 Quantum theory and special relativity – again

Adopting a non-ontic account of quantum states does not address all the challenges raised in the literature concerning the alleged incompatibility between quantum theory and special relativity. According to Bell, for instance, there is 'an apparent incompatibility, at the deepest level, between the two fundamental pillars of contemporary theory' (Bell [2004], p. 172) (where by the 'two fundamental pillars' he means quantum theory and relativity theory), which is not merely due to the difficulty of reconciling collapse with relativity. This difficulty, as Bell acknowledges, is rather easily avoided by 'not grant[ing] beable status to the wave function' (Bell [2004], p. 53; for more on Bell's notion 'beable' see Section 10.2), or, in other words, by adopting a non-ontic account of quantum states.

As Bell and others see it, there are much more profound reasons for believing that quantum theory and special relativity are in conflict than merely the instantaneous and manifestly non-local character of collapse, and these are all related to the fact that the correlations predicted on the basis of entangled states violate a criterion due to Bell that he introduces as *local causality*. This criterion is a technically precise probabilistic implementation of the intuitive idea that in a theory that is compatible with the space-time structure of special relativity, as Bell puts it, '[t]he direct causes (and effects) of events are near by, and even the indirect causes (and effects) are no further away than permitted by the velocity of light' (Bell [2004], p. 239).

Bell's own famous *theorem*[64] states that any theory which respects Bell's probabilistic formulation of local causality cannot possibly reproduce the quantum correlations. Thus, according to this theorem, neither quantum theory itself nor any (possibly more fundamental future) theory which reproduces its predictions can be *locally causal* in the sense of that criterion. Since the quantum theoretical predictions have been spectacularly confirmed in experiments so far, what the theorem seems to show is that *any* candidate theory of fundamental physics, by necessarily being 'non-local', will exhibit the same serious tension with special relativity as quantum theory.

The worry concerning the compatibility between quantum theory and special relativity that arises from the violation of Bell's criterion of local causality is highlighted and investigated in great detail in Maudlin's seminal book-length investigation (Maudlin [2011]) of the topic, which, in the end, arrives at a rather pessimistic conclusion as regards the compatibility question. The central claims of Maudlin's investigation are endorsed in a recent *Scientific American* article (Albert and Galchen [2009]), whose authors conclude that there is indeed a 'quantum threat to special relativity', as announced in the article title. The same conclusion is reached in a lucid reconstruction of Bell's original argument by Travis Norsen, who urges physicists to 'appreciate that there really is here a serious inconsistency to worry about' (Norsen [2011], p. 293). In a similar vein, Michael Seevinck contends 'that a good and fair case can be made that a basic inconsistency exists between quantum theory and relativity' (Seevinck [2010], p. 4). As these passages show, the once common view that quantum theory and special relativity are perfectly able to 'peacefully coexist' (Shimony [1978]) is nowadays under serious pressure. As a matter of fact, it has even been abandoned by Shimony, the one who historically coined the expression 'peaceful coexistence'.[65]

The aim of the present chapter is to argue, contrary to the authors just quoted, that, properly construed, local causality is by no means violated in quantum theory. To support this claim, I will show that Bell's probabilistic criterion of local causality fails to conform to the very natural and highly suggestive idea – already mentioned in Section 6.2.3 while discussing David Lewis' *Principal Principle* – that objective probability ('chance') can only be what imposes constraints on rational degrees of belief. As I will argue, if one accepts the Lewisian connection between objective probability and rational credence as constitutive for what counts as objective probability, the criterion of 'no superluminal signalling', which physicists normally regard as implementing the requirements of relativistic space-time structure (for more details see Section 10.5), is a much better candidate for an adequate implementation of local causality than the probabilistic criterion suggested by Bell. So, we have no reason for thinking that quantum theory really violates local causality, if spelled out properly, and, therefore, as far as Bell's criterion is concerned, no reason to be worried that quantum theory might not be compatible with the space-time structure of special relativity. The structure of the remaining sections of this chapter is as follows: Section 10.2 reviews the discussion of whether quantum theory permits superluminal causation according to a counterfactual analysis of causation and then introduces the details of Bell's probabilistic implementation of local causality and its motivation. Section 10.3 turns to the Principal Principle and recapitulates the role of the variable denoting 'admissible evidence' in its formulation. Bell's criterion of local causality is critically re-examined in Section 10.4 in the light of the Principal Principle. Section 10.5 considers whether the condition of 'local commutativity', which physicists normally take to implement the causal structure of relativistic space-time, fares better than Bell's criterion and comes to a tentatively positive conclusion. The charge of anthropocentrism is taken up again in Section 10.6.

10.2 Formulating local causality

Bell's intuitive characterisation of local causality equates it with the absence of superluminal causal influences. To recapitulate, for him a theory is intuitively locally causal if it fulfils the following criterion (where, to introduce my own label, '(ILC)' stands for 'intuitive local causality'):

> (ILC) The direct causes (and effects) of events are near by, and even the indirect causes (and effects) are no further away than permitted by the velocity of light. (Bell [2004], p. 239)

The criterion (ILC) has two aspects: first, that in a locally causal theory an event and its 'direct' causes and effects are spatio-temporally close to each other, which means that the transmission of causal influences must somehow be 'gapless'; and, second, that in a locally causal theory the transmission of causal influences may not occur at velocities larger than the velocity of light. The two aspects are related, but conceptually distinct. However, Bell himself and those who have followed and accepted his main line of thought have usually focused on whether or not superluminal causal influences can occur in quantum theory and have ignored the question of whether 'gapless' transmission of causal influences is possible. Here I will follow this custom.

Figure 10.1 depicts two space-time regions 1 and 2 as well as the *backward light cone* of region 1, labelled region 3, from which influences

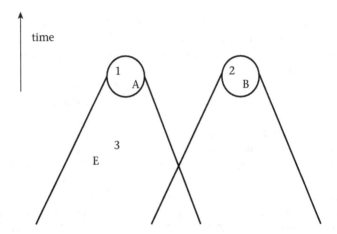

Figure 10.1   Space-time structure of special relativity with space-like separated regions 1 and 2 and associated localised events *A* and *B*

travelling at velocities *at* or *below* the velocity of light can reach region 1. Using this terminology, we can reformulate (ILC) by saying that a theory *T* is locally causal if and only if, according to it, an event can only be causally influenced by events which are not space-like separated from it. Furthermore, I will follow Bell in his application of the criterion (ILC) by assuming that it tacitly rules out causes in the future light cone, i.e. that it defines a theory as locally causal theory if, according to it, the causes of events lie all in their *backward*, rather than forward, light cones.

The main reason why superluminal causal influences are widely deemed problematic in the context of special relativity is that they are *backwards in time* in some inertial frames. Bell himself motivates his insistence on the importance of local causality by highlighting this point. Alluding to the paradoxical features commonly associated with backward causation, he suggests that '[t]o avoid causal chains going backward in time in some frames of reference, we require them to go slower than light in any frame of reference' (Bell [2004], p. 236). However, even though it is true that local causality (in the sense of (ILC)) appears to be naturally motivated from the space-time structure of special relativity in the eyes of many physicists, there is no (and arguably can be no) knock-down proof that superluminal causal influences are impossible in a theory which respects the space-time structure of special relativity.[66] Moreover, as Huw Price convincingly argues, causal influences may even be 'backwards in time' in all inertial frames (i.e. propagate *subluminally* with an inverse 'sign'), without by themselves being in any way incompatible with special relativity.[67]

10.2.1 Causation and counterfactuals

The standard modern approach to the analysis of causation is in terms of counterfactuals. This approach was famously championed by David Lewis (Lewis [1973]) and has since been defended and revised in many forms. The essential idea here is that an event *A* is legitimately regarded as the *cause* of an event *B* just in case, had *A* not occurred, *B* would not have occurred. Several philosophers[68] have argued that on a counterfactual approach to causation quantum theory does indeed sanction superluminal causal influences. In what follows I give an outline of why, based on an epistemic account of quantum states such as the Rule Perspective, it seems natural to deny that quantum correlations require superluminal causation on a counterfactual account of causation. Bell's probabilistic criterion of local causality is discussed in the following subsection.

In his application of the counterfactual approach to causation to the question of superluminal causation Maudlin offers the following *sufficient* condition for superluminal influences, where 'CCLC' (my terminology) stands for 'counterfactual condition of local causality':

(CCLC) If no causal influences are superluminal then it *cannot* be the case for spacelike separated *A* and *B* that *A* would not have occurred had *B* not occurred *and everything in A's past light cone been the same*. (Maudlin [2011], p. 118. The emphasis is Maudlin's)

The clause which adds that everything in *A*'s light cone must be kept fixed in order for there to be genuinely superluminal influences is necessary in order to avoid the mistaken classification of effects of a joint common cause (a so-called screener-off) as causing each other.

From the point of view of an epistemic account of quantum states, applying the criterion (CCLC) to quantum theory is not completely straightforward: according to such an account, there is no such thing as the true quantum state which describes all that happens in region 3, i.e. *A*'s past light cone. So, unless other variables are specified in terms of which candidate full descriptions of space-time region 3 can be given, it is unclear whether there is any such thing as a totality of facts about *A*'s past light cone, so it need not be taken for granted that the expression 'everything in *A*'s past light cone' has a well-defined extension at all.

However, while the Rule Perspective is not meant to *rely on* the hypothesis that a complete description of A's backward light cone can be given, it is not meant to rule out that NQMCs may be used to achieve this. I will therefore grant the assumption that we have a sufficiently rich, perhaps even complete, description of region 3 to enable the application of the criterion (CCLC).

Granting this assumption, that (CCLC) does not hold in quantum theory may seem evident: just consider a situation where both Alice and Bob measure the same spin direction for particles prepared in an EPRB pair with $|\psi_{EPRB}\rangle = \frac{1}{\sqrt{2}}(|+\rangle_1|-\rangle_2 - |-\rangle_1|+\rangle_2)$ (see Eq. (2.19)). Assuming that Alice obtains the result $+1/2$ and Bob obtains the result $-1/2$ and evaluating the Born Rule probabilities derived from $|\psi_{EPRB}\rangle$, it seems clear Bob would not have obtained $-1/2$ had Alice not obtained $+1/2$ with the past light cones of both regions 1 and 2 held completely fixed. Alice's outcome seems to depend counterfactually on Bob's, and conversely, and the criterion (CCLC) seems to be straightforwardly violated. So, by this criterion, quantum theory does indeed seem to sanction superluminal causal influences.

This conclusion, however, depends heavily on a specific *interpretation* of the counterfactual used to evaluate the principle (CCLC). Maudlin himself regards this move as unproblematic on grounds that, as he claims, '[c]ounterfactual claims can be implied by laws if the antecedent of the conditional is precisely enough specified' (Maudlin [2011], p. 120). According to him, it follows from the 'laws' of quantum theory, formulated using quantum states, that the criterion (CCLC) is violated in quantum theory. On an epistemic account of quantum states, however, it is unclear what those 'laws' might be; the Schrödinger equation, for example, is interpreted as a rule that produces advice on how to develop the state assigned to a quantum system in time, not as a law of nature which governs that evolution.

In Section 9.2, I discussed how aspects of a causal account of scientific explanation can be incorporated in the Rule Perspective, and I pointed out the *agency theory* of causation by (Menzies and Price [1993]) as an account of causation that may help one to do this. The agency theory belongs to a wider class of so-called *interventionist* accounts of causation all of which have received much attention from philosophers of science in recent years.[69] On an interventionist account of causation, what it takes for X to cause Y, according to a rough and ready characterisation of Woodward, 'is that an intervention on X with respect to Y changes the value of X in such a way that if any change occurs in Y, it occurs only as a result of the change in the value of X and not from any other source' (Woodward [2003], p. 14).

According to an interventionist account of causation, those counterfactuals that are genuinely *causal* are of the form 'If the value of B were to be changed as the result of an intervention, then the value of S would change' (Woodward [2003], p. 15). Since interventionist accounts of causation have a number of independently attractive features, it is natural to investigate the result of applying one to quantum theory in the light of the criterion (CCLC). Does quantum theory sanction the *causal* counterfactual (interpreted along the lines of interventionism) that Alice would not have obtained the result $+1/2$ had Bob not obtained the result $-1/2$?

Giving a positive answer to this question would require making sense of an intervention that acts 'directly' on the measurement outcome, i.e. one that makes sure that 'Bob obtains $-1/2$' becomes true, rather than 'Bob obtains $+1/2$'. As a matter of fact, we do not know any possible interventions which could fix the measurement outcome without affecting the preparation history and thereby altering the quantum state to be assigned.[70] However, since hypothetical interventions should not be required to be *physically* possible anyway,[71] this does not yet allow us to deny that, at least in principle, there *could be* an intervention on the outcome of Bob's measurement which has all the desired properties.

One possible reaction to this challenge is to look for a contrast between what is possible and impossible that is suggested by quantum theory itself. And indeed such a contrast is embedded in how we apply quantum theory, namely in that we regard it as *possible* to intervene in the system by deciding which observable to measure and as impossible to (deterministically) fix the outcome beyond the preparation of a quantum state. By this standard, there is no intervention by means of which Alice could possibly make it the case that the outcome of her measurement is, say, $-1/2$, and the counterfactual 'Had Alice not obtained the result $+1/2$, Bob would not have obtained the result $-1/2$' does not hold. Thus, quantum theory, on this reading, does *not* sanction superluminal causal influences. But what happens on a wider account of which interventions are possible?

Healey offers an argument, based on Woodward's interventionist account of causation, that outcome-fixing interventions are impossible in the quantum context on *conceptual* grounds.[72] The argument can be summarised as follows:

First, suppose (for reductio) that there is a possible intervention I_1 which fixes Alice's outcome to the value $+1/2$. Second, suppose that, as symmetry considerations suggest, in the case of space-like separated measurement events on Alice's and Bob's sides, there is an analogous possible intervention I_2 which fixes Bob's outcome to the value $+1/2$ (or $-1/2$). Third, note that, if we consider counterfactuals that are grounded in the EPRB state and claim perfect anti-correlations between the results on Alice's and Bob's sides for the same measured observable on both sides, this entails that both I_1 and I_2 can individually be used to fix the outcomes on Alice's and Bob's sides to specific values. Fourth, observe that this is incompatible with the idea that each intervention, by itself, fully controls the outcome it is designed to set to a specific value, for an intervention on the distant system could always be able to disrupt this control in at least one direction. Thus, by reductio, interventions I_1 and I_2 as postulated by this argument are impossible, and, as a consequence, the counterfactual 'Had Alice not obtained the result $+1/2$, Bob would not have obtained the result $-1/2$' does not hold.

That this counterfactual does not hold on Woodward's interventionist analysis depends on his necessary condition for I to be an intervention variable with respect to X that, for appropriate values of I, 'X ceases to depend on the values of other variables that cause X and instead depends only on the value taken by I' (Woodward [2003], p. 98, condition (**IV**) I2). By Healey's argument, the symmetry of the EPR-type scenarios precludes complete direct control of measurement outcomes even as a *conceptual* possibility and thus rules out the counterfactual which implies the existence of superluminal causation. So, on the leading interventionist reading of causal counterfactuals, superluminal causal influences do not seem to be supported by quantum theory. Let us continue our investigation of whether quantum theory and special relativity are in conflict by focusing on Bell's *probabilistic* way of spelling out what it means for a theory to be locally causal.

10.2.2   Local causality probabilistically

Bell acknowledges that the intuitive criterion of local causality (ILC) quoted above 'is not yet sufficiently sharp and clean for mathematics' (Bell [2004], p. 239). If we want to decide whether quantum theory with its probabilistic predictions is to be viewed as conforming to (ILC) or not, we need a criterion that expresses (ILC) in terms of probabilities. According to Bell, any probabilistic formulation of (ILC) should be 'viewed with the utmost suspicion' since 'it is precisely in cleaning up intuitive ideas for mathematics that one is likely to throw out the baby with the bathwater' (Bell [2004], p. 239). This is surely correct: as I will argue, Bell's own preferred way of making (ILC) 'sharp and clean for mathematics' is incompatible with the Principal Principle in conjunction with a highly natural intuitive *probabilistic* rendering of (ILC).

Bell's own probabilistic formulation of local causality appeals to what he calls 'local beables'. According to him, 'the *be*ables of the theory are those entities in it which are, at least tentatively, to be taken seriously, as corresponding to something real', and he specifies that '*local* beables are those which are definitely associated with particular space-time regions' (Bell [2004], p. 234). For our present purposes, we might picture the local beables as objective properties or events within 'local' space-time regions. Using 'local beable', Bell's account of what it means for a theory to be locally causal is the following:

A theory will be said to be locally causal if the probabilities attached to the values of local beables in a space-time region 1 are unaltered by specification of values of local beables in a space-like separated region 2, when what happens in the backward light cone of 1 is already sufficiently specified, for example by a full specification of local beables in a space-time region 3.[73] (Bell [2004], pp. 239f.)

This still leaves room for interpretation, but the essential idea is easily identified as the following: if some theory is locally causal, the probability of any event occurring is completely determined by what happens in the backward light cone of the region where it (potentially) occurs; it does *not* depend on what happens at space-like distance (or in the forward light cone). It is useful to attach a label to this (still somewhat vague and intuitive) idea of what it means for a theory to be locally causal and call a theory 'intuitively probabilistically locally causal' (in what follows denoted '(IPLC)') if and only if according to it:

(IPLC)   The probability $Pr(A)$ of an event A in a space-time region 1 depends only on what events occur in the backward light cone of region 1.

Before discussing how Bell himself renders the intuitive probabilistic criterion (IPLC) precise, let us briefly reflect on why one might think that theories must conform to (IPLC) in order to be compatible with special relativity. According to Maudlin, the incompatibility between quantum theory and special relativity that he claims exists arises from the fact that in quantum theory a measurement result's *probability* depends on something at space-like distance ('the distant setting' in the following quote). As he claims, a preferred reference frame is introduced by this 'non-local dependence', which is therefore incompatible with special relativity:

> The underlying problem for a relativistic theory, then, is not wave collapse *per se* but rather the non-local dependence of one measurement result on the distant setting. In collapse theories, that dependence is secured through collapse; in Bohm's theory it is mediated through the uncollapsed wave-function. Wherever there is such a dependence then ... there must also be a cut-off point beyond which, e.g., the setting on the right can no longer have an effect on the left, for the distant measurement can be postponed indefinitely, or never performed at all. But that cut-off point defines a preferred frame of reference, namely, the frame in which the cut-off is simultaneous with the measurement on the left. (Maudlin [2011], p. 196)

For the purposes of the present investigation, I will not contest the claim that (IPLC) (or, if we allow backward influences, a time-symmetric cousin) must hold in a theory for it to be compatible with special relativity. However, to decide whether quantum theory really violates (IPLC), the latter has to be made more precise, as it is ambiguous in two ways: first, it is open to interpretation what one should regard as *the* probability (simpliciter) of an event; second, it is open to interpretation how to specify those parts of space-time on which that probability should be taken to depend.

Bell himself endows the criterion (IPLC) with a more precise mathematical meaning by defining a locally causal theory as one where the probability of any event A in some space-time region 1, *given* a complete specification E of what happens in its backward light – i.e., the probability $Pr(A|E)$ – , may not be changed by specifying anything that happens in some space-like separated region 2. In other words, T is locally causal according to Bell's criterion if, for any event A in region 1, any event B in a space-like separated region 2, and a complete specification E of all that occurs in the backward light cone of region 1, the following equality holds:

(BPLC) $Pr(A|E) = Pr(A|E,B)$,

where '(BPLC)' stands for 'Bellian probabilistic local causality'.[74]

The criterion (BPLC) certainly appears very natural as a way of rendering (IPLC) mathematically 'sharp and clean'.[75] One may try to motivate it along more metaphorical lines as follows: consider region

1 immediately prior to 'deciding' whether A is to happen or not. The outcome, one may suppose, will depend on how the relevant probabilities 'play out' whether A and B. But if nature is locally causal in the sense of (IPLC), the relevant probabilities, in turn, will depend only on what occurs in the backward light cone of region 1. So, the relevant probability $Pr(A)$ which governs the behaviour of the system in region 1 (the probability which, metaphorically speaking, the system 'uses' when 'deciding' whether A is to happen or not) must be given by $Pr(A|E) = Pr(A|E,B)$ in a locally causal theory.

However, even though this metaphorical motivation for (BPLC) may appear natural, it arguably rests on misleading preconceptions concerning the role of time. It is reconsidered in Section 10.6 and rejected for being based on an inconsistent partial anthropocentrism.

The reasons for thinking that quantum theory violates (BPLC) are essentially the same as those for thinking that it violates (CCLC). Ignoring any difficulties associated with the idea of a complete description of A's backward light cone mentioned already in the previous section, we can let the role of the variable E be played by ψ_{EPRB}, which ascribes perfectly anti-correlated results for measurements of, say, spin-1/2 particles in regions 1 and 2 where the same spin direction is measured for both systems.[76]

Then, if we let A correspond to the event 'measurement of spin in z-direction in region 1 yields 1/2' and B correspond to the event 'measurement of spin in z-direction in region 2 yields 1/2', we obtain

$$Pr(A|\psi_{EPRB}) = 1/2 \tag{10.1}$$

and

$$Pr(A|\psi_{EPRB},B) = 0. \tag{10.2}$$

This manifestly contradicts (BPLC) and establishes that the existence (and preparability) of entangled quantum states such as ψ_{EPRB} implies the violation of (BPLC) in quantum theory. This result is strengthened by Bell's theorem, which states that no (potentially more fundamental) theory that recovers correlations as predicted by quantum theory can possibly respect (BPLC). Since quantum theory is extremely well confirmed as far as its predictions based on entangled states are concerned, one naturally concludes that (BPLC) is violated in nature. Thus, if (BPLC) is indeed the appropriate mathematically precise rendering of (IPLC) and the latter is thus also violated in nature, then, if Maudlin (interpreted as above) is correct, special relativity is indeed threatened by quantum theory.

However, as I shall argue in what follows, by the standards of the Principal Principle, (BPLC) does not correctly implement the criterion (IPLC). If this is right, then the fact that (BPLC) is violated does not show that (IPLC) is violated as well. And indeed, as I am going to argue later, we have good reason to believe that (IPLC) holds in quantum theory even if (BPLC) does not. Consequently, even if we accept the claim that violations of (IPLC) in quantum theory *would* give rise to an incompatibility with special relativity, they would be harmless because (IPLC) is best *not* taken to be violated in quantum theory.

In what follows, I review Lewis' 'subjectivist guide' (Lewis [1986a]) to objective probability and the Principal Principle as its core element, which will form the basis of my argument for this claim.

10.3 The Principal Principle and admissible evidence

To begin with, let us assume that, as argued for in Section 6.2, the probabilities dealt with in quantum theory are in some sense *objective*. And indeed, according to the Rule Perspective, they *are* manifestly objective inasmuch as their ascriptions are governed by objective standards of correctness. In Section 6.2.3, the Principal Principle was already recommended as the defining link between objective probabilities and rational degrees of belief. Its main motivation, in Lewis' own words, is the following:

> Don't call any alleged feature of reality 'chance' unless you've already shown that you have something, knowledge of which could constrain rational credence. (Lewis [1994], p. 484)

Let us now formulate the Principal Principle in slightly more formal terms. It states that, for any coherent prior credence function $cr()$[77] and any arbitrary proposition A concerning some chancy matter of fact,[78] the quantity x deserves to be called the 'objective probability' of A – $Pr(A)$, for short – only if (where '(PP)' stands for 'Principal Principle')

(PP) $cr\left(A|Pr(A) = x, E\right) = x.$

The variable E stands for 'admissible evidence' (Lewis [1986a], pp. 92–96) and will play an important role in what follows.[79] Read in the other direction the Principal Principle states that, if one knows that x is the objective probability of A and one only has 'admissible evidence', then one's rational degree of belief concerning A is x.

If an agent has evidence that allows her to have more informed expectations regarding A than she could have by knowing A's chance, she is said to have 'inadmissible' evidence, i.e. evidence that goes beyond the 'admissible' evidence E in (PP). When one tries to put the Principal Principle to work by extracting the rational credences from the chances, if known, or the chances from the rational credences, if known, the challenge essentially amounts to determining the admissible evidence: given the chances, the rational credences are fixed only if the admissible evidence is specified, and given the rational credences, the chances are fixed only if the admissible evidence is specified. The admissible evidence relative to some agent (or agent-position) must always contain the evidence about the chance $Pr(A)$ relative to that position itself: evidence can be *inadmissible* only if having it makes it potentially rational *not* to align one's credence about A with A's objective probability. It follows that evidence as regards A's objective probability itself is always admissible.

Agents' rational credences with respect to one and the same proposition A may differ if they have access to different bits of evidence when forming their credences. In David Lewis' account, the insight that rational credences (and, consequently, total amounts of admissible evidence) are different for different agents is reflected in taking total amounts of admissible evidence to be relative with respect to the *time*, parameter t. Given the Principal Principle, treating total amounts of admissible evidence as relative with respect to t is equivalent to treating objective probabilities themselves as relative with respect to t. Lewis' 'reformulation' of the Principal Principle using a 'chance theory' T and the complete history H_t of the world up to the time t makes this explicit:

(TPP) $cr(A|H_t, T) = Pr_t(A),$[80]

where '(TPP)' has been introduced for 'time-relativised Principal Principle' and objective probability $Pr_t(A)$ is explicitly time-relative. On Lewis' account, evidence about the past is in general admissible, whereas evidence about the future is not. In the special relativistic space-time framework this creates problems, however, as it does not contain any universally preferred simultaneity relation and hence no preferred way of singling out the time-parameter t in (TPP). Special relativity hosts no preferred way to carve up space-time into past and future, so the formulation (TPP) of the Principal Principle and the associated time-relativisation of objective probabilities are of no use for our present purposes.

The most obvious way out of this difficulty may appear to be taking chances to be relative to space-like *hyperplanes*, i.e. planes of constant time t in a specific inertial frame. This is proposed by Wayne Myrvold, who proposes it as a move towards making sense of 'relativistic quantum becoming' (the title of (Myrvold [2002])). Myrvold correctly observes that 'it is not nonsensical to conditionalise on events not in the past' (Myrvold [2002], p. 492) (or, in the relativistic context, outside one's backward light cone) since quantities $Pr_{E_h}(A)$ which are obtained by conditionalising $Pr()$ on everything on the 'backward' side of the space-like hyperplane h may be well-defined. To observe that such quantities can be formed does not yet establish, however, that they deserve to be called 'objective probabilities' in the sense of the Principal Principle. And indeed, it turns out that the quantities $Pr_{E_h}(A)$ cannot be the objective probabilities in the sense of the Principal Principle, for they do not yield the agents' rational credences.

To see this, it suffices to observe that the space-like hyperplanes are indefinitely stretched out in space-time; consequently, if $Pr_{E_h}(A)$ specifies the rational credence of an agent on earth with respect to A, it should also be the rational credence for some (hypothetical) agent located myriads of light years away in the cosmos with respect to A. But concerning matters of fact on earth, the rational credences of the one located here and the one who is myriads of light years away cannot be the same: the amount of information they can access as to what occurs here on earth is very different, so their rational credences about *some* contingent matters of fact are clearly different.

Unlike agents who are located along the same space-like hyperplane, agents who are located in the same (finite and not overly large) space-time region may at least in principle share all the information they have as regards what events have occurred (or will occur) across space-time. So, if we accept this modest and arguably very natural degree of idealisation that seems to go rather well with the Rule Perspective, it makes sense to assume that rational credences – and so, by the Principal Principle, the objective probabilities themselves – are relative to (by some reasonable measure) finite and not overly large space-time regions.

In the previous section we already encountered space-time regions that are finite and not overly large when we considered the space-like separated space-time regions 1 and 2 and events A and B that were (potentially) occurring within them. As it is natural to take rational credences to be relative to such regions, it is natural, in view of the Principal Principle, to take the objective probabilities to be relative to the same regions in the context of special relativistic space-time as well.

One may ask about the exact size and form of appropriate regions and whether one might choose them to be point-like, but I wish to neglect these questions here, assuming that they can be settled in a satisfying way. After introducing the variable y to range over space-time regions of the appropriate nature and shape the Principal Principle takes the form (where '(RPP)' stands for 'region-relativised Principal Principle'):

(RPP) $\qquad cr(A|E_y, T) = Pr_y(A)$.

Here again 'T' denotes the theory which we use to compute the probabilities and 'E_y' now denotes the total amount of admissible evidence relative to space-time region y. The admissible evidence E_y will be focused on in the next section, where I use the formulation (RPP) to look for an implementation of the intuitive probabilistic principle of local causality (IPLC) that conforms to the Principal Principle.

10.4 Intuitive probabilistic local causality in the language of the Principal Principle

The Principal Principle specifies agents' *rational* credences using objective probabilities and, as such, is a *normative* principle. As a normative principle, it must be seen in the light of one of the most important maxims of practical philosophy, namely, 'ought implies can'. This maxim plays a central role when one wishes to determine what credences agents *should have* in the quantum theoretical context by linking it to what evidence agents possibly *can have*, given how they are spatio-temporally situated.

More specifically, if the theory T that is used by an agent to compute the probabilities guiding her credences rules out that she can have any evidence as to whether B (for some chancy local event B), then evidence that B has occurred cannot possibly be admissible for her, according to T. If our agent has a 'hunch' that B occurs, while T excludes that she can have any evidence as to whether B, it would clearly not be rational for her to form her credences by conditionalising over B. This holds even if conditionalising over B would enhance her predictive and pragmatic success. If, however miraculously, it 'occurs' to our agent that B (somehow 'magically'), while she is perfectly aware that, if T is correct, she cannot have any evidence as to whether B, it is clearly irrational for her to conditionalise over B. (She might well be described as a 'lucky fool' if she did). This thought crucially underlies the present argument that Bell's criterion (BPLC) does not adequately spell out the intuition captured in (IPLC) if we assume the Principal Principle.

The argument itself starts with an implementation of the intuitive probabilistic criterion (IPLC) that respects the Principal Principle in the form (RPP). In order to implement the criterion (IPLC), we must first decide which value of $Pr_y(A)$ (i.e. the value of Pr_y for which value of the index y) should be taken to be *the* probability of A in the sense of (IPLC). The only non-arbitrary choice is $Pr_1(A)$, i.e. the probability of A relative to region 1, where it (potentially) occurs. By the Principal Principle, the probability $Pr_1(A)$ is determined, given the correct 'chance theory' T, by the total amount of admissible evidence E_1 relative to region 1. Let us decompose E_1 as the conjunction of the total admissible evidence E_{1-} concerning the backward light cone of 1 and the total admissible evidence $E_{1,0}$ concerning what happens *outside* the backward light cone: $E_1 = E_{1-} \wedge E_{1,0}$. The principle (IPLC) which we want to implement in a way that conforms to the lesson of the Principal Principle mandates that, for T to be locally causal, $Pr_1(A)$ may depend only on what happens in the backward light cone of region 1, i.e. only on E_{1-}. In other words, the principle (IPLC) mandates that the probability $Pr_1(A)$, which, recall, is fully determined by the complete admissible evidence E_1, i.e. $Pr_1(A) = Pr_1(A|E_1)$, must be equal to the probability $Pr_1(A|E_{1-})$, where E_{1-} is the complete admissible evidence about the backward light cone of region 1:

$$Pr_1(A) = Pr_1(A|E_{1-}). \tag{10.3}$$

This is the appropriate formal expression of what it takes for a theory to be locally causal by the standards of (IPLC) in the light of the Principal Principle. We can now compare Eq. (10.3) and Bell's criterion (BPLC) and see whether they yield the same verdicts as to which theories are locally causal.

Even though Eq. (10.3) and (BPLC) look very similar, differences may arise since the latter does not pay attention to admissibility. To begin with, assume that, relative to region 1, there is no inadmissible evidence about the backward light cone of region 1 and no admissible evidence about what happens in its future light cone. Further, let us introduce a variable, I_1, for the complete inadmissible evidence about what occurs at space-like separation from region 1. Bell's criterion (BPLC), which states that $Pr(A|E, B) = Pr(A|E)$ for arbitrary events B at space-like separation from region 1, whether admissible or not, then translates into the statement

$$Pr_1(A|E_1 \wedge I_1) = Pr_1(A, E_{1-}). \tag{10.4}$$

This is *not* equivalent under all circumstances with Eq. (10.3), particularly not if I_1 is non-empty and, when conditionalised on, has a non-trivial impact on Pr_1. If, for some theory T, the complete admissible evidence E_1 with respect to region 1 reduces to evidence about the backward light cone – i.e. if E_1, T and E_{1-}, T are equivalent – then, by the Principal Principle, Eq. (10.3) holds trivially for the theory T, whereas Eq. (10.4) and, thus, Bell's criterion (BPLC) may well be violated due to the relevance of I_1. Returning to quantum theory, while it violates Eq. (10.4), considerations on the rational credences of competent users of it strongly suggest that it does conform to Eq. (10.3), as I argue in the following section.

10.5 Quantum theory, local causality, and the Principal Principle

Quantum field theorists typically take the assumption that it is impossible to causally influence what occurs at space-like distance to be implemented by the principle that the operators representing observables that are associated with space-like separated regions (or space-time points) must commute. Formally, if $A(x)$ and $B(y)$ are operators representing observables that are associated with the space-like separated points x and y, this principle says:[81]

$$[A(x), B(y)] = A(x)B(y) - B(y)A(x) = 0. \qquad (10.5)$$

In physical terms, Eq. (10.5) means that measuring $A(x)$ at (or close to) the point x is statistically irrelevant for the outcomes of measuring $B(y)$ at (or close to) the space-like separated point y. As a consequence, an agent who is located near x cannot *send any signal* to an agent who is located at y by the choice of observable $A(x)$ she decides to measure. In its mathematically precise form this statement is widely known as the 'no- (superluminal) signalling theorem'.[82] Since physicists widely regard the impossibility of sending superluminal signals as the appropriate implementation of the relativistic space-time structure, they often refer to Eq. (10.5) as '*relativistic causality*' (or '*causality*'). (Haag [1993], p. 57) But how does Eq. (10.5) – and the associated impossibility of superluminal signalling – fare as a way of spelling out the intuitive probabilistic criterion (IPLC) in the light of the Principal Principle?

There is one obvious difference between Eq. (10.5) and (BPLC), namely that Eq. (10.5) is time-symmetric whereas (BPLC) (just as (IPLC) is) is not. As a consequence, Eq. (10.5) can at best be the formal implementation of a time-symmetric cousin of (IPLC) that refers to the backward and forward light cones of region 1 taken together and not merely to the backward light cone. As long as we ignore the thorny issue of backward probabilistic dependences, however, we can still ask whether Eq. (10.5) is successful in capturing the part of (IPLC) that excludes probabilistic dependencies between *space-like separated* regions.

The impossibility of superluminal signalling seems to be *necessary* for (IPLC) (or, more precisely, for its time-symmetric cousin): if superluminal signals were possible, evidence concerning events that occur at space-like separation and cannot be predicted from what occurs in the backward light cone along would in principle be accessible (via superluminal signals), so rational agents would in general have excuse for neglecting it as the basis for their credences. Evidence concerning what occurs at space-like separation would then be accessible and, hence, most plausibly *admissible* as well as, by the Principal Principle, having non-trivial impact on the objective probabilities. This implies that neither (IPLC) nor its time-symmetric cousin can hold in a theory which allows superluminal signalling.

In a theory where, in contrast, superluminal signalling is *impossible*, it does not seem to be possible for an agent to obtain information regarding what happens at space-like separation from herself unless she can derive it from evidence as regards what occurs in her own backward light cone (or total light cone, if we allow backward causation) by the help of the theory T which she uses. This suggests that in such a theory all accessible – and, so, all admissible – evidence is about what occurs in the backward (or total) light cone, and the implementation Eq. (10.3) of (IPLC) is fulfilled.

However, not every theory which incorporates Eq. (10.5) and respects the principle of no- (superluminal) signalling is compatible with special relativity. It may still rely on such notions as absolute simultaneity and a preferred reference frame, thus defying a fully relativistic formulation. In fact, this is precisely the reason why interpretations or modifications of quantum theory such as pilot wave theory and GRW theory (in their standard formulations) are not easily formulated in a relativistic setting. According to Maudlin as quoted in Section 10.2.2, these difficulties are inevitable in relativistic theories which reproduce the quantum theoretical predictions due to what he calls 'the non-local dependence of one measurement result on the distant setting' (Maudlin [2011], p. 196). From the point of view of the present considerations, in contrast, a more optimistic perspective is available: if we trust the Principal Principle, we have no reason to suppose that the relevant quantum probabilities depend on events at space-like distance and, so, inasmuch as this worry is concerned, no reason to suppose that quantum theory by *itself* gives rise to an unavoidable conflict with special relativity.

10.6 But how does nature perform the trick?

Critics may object to the previous considerations that quantum theory cannot possibly fulfil Eq. (10.3) by arguing that quantum theoretical probabilities cannot depend only on what occurs in the backward light cone. If they did, the outcome statistics in space-like separated regions would be *independent*, which they manifestly are not.

The criticism can be illustrated using the EPR-Bohm state ψ_{EPRB} according to which measurement results on the wings of an experiment carried out in regions 1 and 2 are perfectly anti-correlated for measurement of the same observable. Let us assume that the same observable (say, 'spin in z-direction') is measured on both (space-like separated) wings for many pairs of systems, all prepared in such a way that ψ_{EPRB} is to be assigned. Then, according to the presented before, the objective probabilities of possible outcomes $+1/2$ and $-1/2$ in regions 1 and 2 are $P_1(\pm 1/2) = P_2(\pm 1/2) = 1/2$ relative to these regions themselves. But, the criticism complains, this would mean that the outcomes are *uncorrelated* and the same result were obtained on both wings in approximately half of the runs of the experiment. This is clearly not what quantum theory predicts. So, the criticism continues, at least one of the probabilities $P_1(1/2)$ or $P_2(1/2)$ must be 0 or 1, i.e. depend on what happens on the other wing of the experiment. Otherwise, the criticism concludes, the pattern of events that quantum theory predicts could not possibly be realised.

In response to this criticism, one may first note that nowhere here has it been denied that the correct *conditional* probabilities $P_1(A|B)$ and $P_2(B|A)$, which are relevant for the rational credences of agents in regions 1 and 2, must be 0 or 1 (in the EPRB example with the same observable being measured on both wings of the experiment). The pattern of events that is to be predicted on the grounds of these probabilities does exhibit the perfect anti-correlations (and whatever other correlations are predicted based on the entangled states). In the EPRB example, a pattern in which sometimes the same result occurs on both wings in the same experimental run is excluded. And why should one require that one of the *unconditional* probabilities $P_1(A)$ or $P_2(B)$, in addition to the conditional ones, be 0 or 1 as well?

The proponent of the criticisms may answer by appealing to what may appear to be an essential ingredient of a thoroughly non-anthropocentric perspective on quantum probabilities: according to this perspective, the genuinely *unconditional* probabilities are the ones which *govern* (or *determine*) how the pattern of events evolves in time. And unless one of *those* is 0 or 1, the critic may hold, nature cannot possibly manage to 'perform the trick'[83] to produce the correlations that quantum theory predicts. Focusing on the probabilities which are relevant for 'nature herself' when she evolves the pattern of events such that it displays the quantum correlations vindicates Bell's criterion (BPLC) as to how (IPLC) should be implemented. Local causality in the sense of (IPLC), according to this criticism, turns out to be violated in the end by the quantum theoretical predictions after all.

Arguably, this criticism rests on a picture concerning the role of time and how the pattern of events 'evolves' in time which 'holds us captive' (to use a Wittgensteinian phrase, (Wittgenstein [1958] § 115)) and distorts our view. This is not the place to review the general arguments against the idea that time passes in an objective sense and that the pattern of events, accordingly, *evolves* in time in an objective sense, which patently form the basis of the criticism just outlined. It suffices to note for our present purposes that the alleged problem cannot be formulated from the perspective which is widely regarded as the *least* anthropocentric view of space-time: the so-called 'block universe' view according to which, in the words of Huw Price, 'reality [i]s a single entity of which time is an ingredient, rather than ... a changeable entity set *in* time'.[84] Thus, according to the block universe view, there is neither an objective 'flow of time' nor an objective sense of 'becoming' or 'coming into being' of events. Everyday ideas such as that of a 'passage of time' or of things 'evolving in time' may still be regarded as perfectly legitimate in the everyday contexts where we normally use them, but they are arguably absent from the most consistently non-anthropocentric perspective on space-time that we can avail ourselves of.

The block universe view is relevant for our present purposes inasmuch as it undermines the idea that there is any 'evolution' of the pattern of events in some objective sense. There is, if we accept it, no question concerning how nature manages to 'perform the trick' of evolving the pattern of events such that the 'emerging' correlations violate (BPLC). From the point of view of the block universe, the appeal to those probabilities which *nature herself* (or the physical systems themselves) must use when evolving the pattern of events in time is based on an inadvertent anthropocentric stance vis-à-vis the role of time. Since the main motivation behind the criticism outlined above was to *avoid* the more overt anthropocentrism of the present analysis with its appeal to the Principal Principle, this result is ironic.

Another way to put this is by saying that, from the point of view of the block universe, the idea that probabilities must depend on events at space-like separation to account for correlations which violate (BPLC) is based on a double standard as regards the role of time and the role of the spatial dimensions.[85] Once this double standard is recognised, it loses its appeal, and the attraction of (BPLC) as a candidate criterion of local causality loses its grip.

To conclude, we can either be consistent in avoiding any appeal to anthropocentric notions whatsoever: then there is no question as to why the pattern of events *evolves* in time as it does, only if it conforms to the quantum theoretical predictions (where 'prediction' is to be understood in an atemporal sense); or we can be consistent in our appeal to (partially) anthropocentric notions (such as, arguably, 'causing', 'bringing about', and 'becoming') and then base our investigation of whether quantum correlations require probabilities which depend on what happens at space-like distance on an analysis that duly respects the anthropocentric features of these notions. The Principal Principle seems ideally suited as the conceptual tool for pursuing this latter goal. The result of applying it, as I have argued, is that we do not have any reason for doubting that quantum theory fulfils local causality, properly construed, and that the purported conflict between quantum theory and special relativity does not exist.

11
A Look at Quantum Field Theory

The therapeutic approach to quantum theory developed here is meant to be applicable not only to non-relativistic quantum mechanics, but to more advanced quantum theories as well. More specifically, it is meant to be applicable to the quantum theories of systems with infinitely many degrees of freedom, quantum field theories in particular. However, the quantum theories of systems with infinitely many degrees of freedom harbour additional philosophical challenges besides those that are already present in non-relativistic quantum mechanics, and it is to be hoped that the approach suggested here may help to clarify and address those challenges as well. The present chapter turns to some aspects of the interpretation of quantum field theory and assesses them in the light of the Rule Perspective.

11.1 Lagrangian versus algebraic quantum field theory

Philosophers approaching quantum field theory typically have to make a decision as to which *type* of quantum field theory they would like to turn to. In particular, they have to decide between conventional 'Lagrangian' quantum field theory, i.e. quantum field theory as actually used by working physicists in high-energy particle physics, and the more mathematically minded *algebraic* approach, which allows much greater mathematical rigour but has so far been without much contact with experimental results. Regrettably, no formulation of interacting quantum field theories in four space-time dimensions has yet been found which is both mathematically rigorous and empirically successful. Even though the Lagrangian and the algebraic approaches are not practised in complete isolation from each other,[86] it is difficult for philosophers *not* to choose either of them as the point of departure from which to start their investigations.

Recent years have seen a spirited controversy as to whether philosophers should engage with the Lagrangian or with the algebraic approach.[87] I feel free not to take sides in this dispute and accept both approaches as worthy of philosophical attention. Nevertheless, my focus will be mostly on the algebraic approach in this chapter, not because I find it intrinsically more interesting or respectable, but because the challenges to which it gives rise seem to be more radically new than those raised by the Lagrangian approach.

The Lagrangian approach retains the conceptual apparatus of ordinary non-relativistic quantum mechanics, starting with canonical commutation relations and a Hilbert space structure and pure states as (equivalence classes of) vectors in this setting. Perhaps the main challenge which faces this approach concerns the nature and legitimacy of the *renormalisation methods* used to extract meaningful physical results from this framework.[88] Further challenges concern, for instance, the status and interpretation of gauge symmetries and their breaking, and this topic will be briefly touched upon further below. None of these challenges seems to be closely linked to which type of account of quantum states one accepts, and it seems therefore acceptable to neglect them for our present purposes.

The challenges raised by the algebraic approach, in contrast, are directly related to the question of how to interpret quantum states. Arguably, taking them seriously provides further motivation for adopting a non-ontic account of quantum states, such as the Rule Perspective. In what follows I substantiate this claim by reviewing some basic features of the algebraic approach and investigating some of the challenges that arise from them in a little more detail.

11.2   Basics of the algebraic approach

Let me start by giving a brief outline of the algebraic approach. To begin with, it is useful to recall that, as explained in Section 2.1, in ordinary quantum mechanics the observables of a quantum system are represented by the self-adjoint elements of an algebra $B(\mathcal{H})$ of bounded linear operators on a Hilbert space $\mathcal{H}$, and the possible states of the system are expressed as density matrices on $\mathcal{H}$. In the algebraic approach, in contrast, one starts in a more abstract way in terms of a C^* algebra[89] $\mathcal{A}$ and defines the quantum states ω as linear functionals from $\mathcal{A}$ into the set $\mathbb{C}$ of complex numbers, such that $\omega(I) = 1$ and $\omega(A^*A) \geq 1$ for all $A \in \mathcal{A}$. To a first approximation (at least), one may think of the self-adjoint elements of $A \in \mathcal{A}$ as observables of the system and of $\omega(A)$ as the expectation value of A ascribed by the state ω.

A Hilbert space representation π maps the C^* algebra $\mathcal{A}$ into an algebra $\mathcal{B}(\mathcal{H})$ of bounded linear operators on a Hilbert space $\mathcal{H}$ while preserving all algebraic relations. In ordinary non-relativistic quantum mechanics for systems with finitely many degrees of freedom, the image $\pi(\mathcal{A})$ of $\mathcal{A}$ under π (or, more precisely, its so-called strong closure $\pi(\mathcal{A})'^{90}$) is isomorphic to some $\mathcal{B}(\mathcal{H})$. As a consequence, and as already pointed out in Section 2.1 in connection with the Stone–von Neumann theorem, all representations π of $\mathcal{A}$ are *unitarily equivalent* in ordinary quantum mechanics. This means that, for any two representations $(\pi_1, \mathcal{H}_1)$ and $(\pi_2, \mathcal{H}_2)$, there exists a one-to-one norm-preserving linear map $U : \mathcal{H}_1 \mapsto \mathcal{H}_2$ such that $U\pi_1(A)U^{-1} = \pi_2(A)$ for all $A \in \mathcal{A}$. Schrödinger's wave mechanics and Göttingen matrix mechanics can be seen as different Hilbert space representations of the same algebra of observables that is generated by the Heisenberg commutation relations, and they are widely and justly regarded as two formulations of one and the same theory. The choice of representation is a matter of computational convenience rather than physical significance in this case. However, outside the scope of ordinary quantum mechanics, that is, in the realm of quantum theories for systems with infinitely many degrees of freedom – QM_∞, for short – where the algebra $\mathcal{A}$ admits unitarily inequivalent Hilbert space representations, things are far less clear.

11.3 'Pristine' interpretations

For the case of QM_∞, there are essentially two fundamentally different approaches to the interpretation of the concepts just introduced. The first of these, for which (Arageorgis [1995]) introduced the term '*Hilbert Space Conservatism*', adamantly sticks to the conventional perspective of identifying the observables of a quantum theory with the self-adjoint elements of a Hilbert space representation $\pi(\mathcal{A})$ (or, more precisely, with the self-adjoint elements of $\pi(\mathcal{A})'$, the closure of $\pi(\mathcal{A})$ in the strong operator topology; see note 90) and possible states ω of the system with density operators ρ on a Hilbert space $\mathcal{H}$ such that $\omega(A) = \mathrm{Tr}(\rho A)$ for all $A \in \pi(\mathcal{A})$. Doing so, the position makes use of the conceptually fundamental *GNS representation theorem* ('GNS', for Gel'fand, Naimark, and Segal), according to which any state ω on $\mathcal{A}$ defines a unique (up to unitary equivalence) Hilbert space representation $\pi_\omega(\mathcal{A})$ such that for some (cyclic) vector $|\xi_\omega\rangle$ on the Hilbert space $\mathcal{H}$ of $\pi_\omega(\mathcal{A})$ all expectation values $\omega(A)$ can be written as $\langle\xi_\omega|\pi_\omega(A)|\xi_\omega\rangle$ for all $A \in \mathcal{A}$.

Unlike in ordinary QM, not all states ω on $\mathcal{A}$ are in general expressible as density matrices on the Hilbert space of a single state ω's GNS representation. Those which can be expressed in that way for some ω are said to be in ω's *folium*. States ω_1, ω_2 giving rise to unitarily inequivalent GNS representations have different folia. If they are *pure* states, their folia are *disjoint*, which means that no state expressible as a density matrix on the Hilbert space $\mathcal{H}_1$ of the GNS representation of ω_1 can be written as density matrix on the Hilbert space $\mathcal{H}_2$ of the GNS representation of ω_2, and vice versa. The Hilbert Space Conservative regards as physically relevant only those states that are expressible as density matrices ρ in the strong closure of a single state ω's GNS representation $\pi_\omega(\mathcal{A})$, that is, only states in the folium of the actual state of the system, whatever it is.

The Hilbert Space Conservative's most important opponent is referred to as the 'Algebraic Imperialist' in the terminology introduced and made popular by (Arageorgis [1995]). Algebraic Imperialism's central claim is that the abstract C^*-algebra $\mathcal{A}$ itself contains the full physical content of a quantum theory in the sense that its self-adjoint elements correspond directly to the physical magnitudes characterising the system and the states defined on $\mathcal{A}$ are identified with the system's possible states. For the Algebraic Imperialist, Hilbert space representations $\pi(\mathcal{A})$ of the algebra $\mathcal{A}$ are mere computational tools without any independent physical significance.

Algebraic Imperialism and Hilbert Space Conservatism disagree on which states are physically relevant – only those in the folium of a particular state ω according to Hilbert Space Conservatism; all states defined on the algebra $\mathcal{A}$ according to Algebraic Imperialism. Similarly, they disagree on what constitutes an observable – the self-adjoint elements of some $\pi_\omega(\mathcal{A})'$ according to Hilbert Space Conservatism and the self-adjoint elements of $\mathcal{A}$ itself according to Algebraic Imperialism. In addition to these two approaches, Ruetsche introduces a third in her important recent book *Interpreting Quantum Theories* (Ruetsche [2011]), an approach which she calls 'Universalism' (Ruetsche [2011], pp. 146f). Universalism construes the content of a quantum theory in terms of a specific representation, namely, the direct sum of the GNS representations of all states over $\mathcal{A}$.

Rather than taking sides in the dispute between the three interpretive strategies just outlined, Ruetsche criticises them all for not being able to account for the full explanatory force which theories of QM_∞ recognisably have. The common feature of all three which she targets as the origin of the difficulties they run into is that they specify the physical content of a quantum theory prior to 'the messy business of applying the theory in question to individual problems' (Ruetsche [2011], p. 146). By doing so, these interpretations conform to what Ruetsche calls the 'ideal of pristine interpretation'. As she explains, this ideal is grounded in the idea that the content of a theory T be construed as 'invariant under changes in T's applications, [so that] T won't admit different interpretations in different settings' (Ruetsche [2011], p. 13).

Ruetsche argues at great length that the shortcoming of 'pristine' interpretations in accounting for the full explanatory force of theories of QM_∞ is precisely *due to* the fact that they follow the ideal of pristine interpretation. In contrast to this ideal, she develops the conception of an 'unpristine interpretation', based on the idea that the content of a physical theory T may depend on how T is actually put to use. In Ruetsche's own words:

> The doctrine of *unpristine* interpretation allows that the contingent application of theories does not *merely select* among some preconfig-ured set of their contents, but *genuinely alters* their contents. It follows that there can be an *a posteriori*, even a pragmatic dimension to con-tent specification, and that physical possibility is not monolithic but kaleidoscopic. (Ruetsche [2011], p. 147, the emphasis is Ruetsche's)

If some theory T is interpreted in an 'unpristine' way in Ruetsche's sense, this means that what is regarded as physically relevant accord-ing to T is dependent on what specific applications of T one has in mind. In an unpristine interpretation of a quantum theory the con-text of application is taken into account to decide what mathematical objects correspond to (which) physical quantities and which states are physically relevant.

11.4 The Rule Perspective and unpristine interpretations

Ruetsche herself does not seem to consider non-ontic accounts of quan-tum states when discussing whether to adopt a pristine or an unpristine interpretation of quantum theories. However, considering the prospects for no-collapse ontic accounts of quantum states she proposes that those accounts may conceive of the 'collapsed' state that is in practice assigned after measurement as a mere 'appropriate predictive instru-mentality when considering [further] measurements' (Ruetsche [2011],

p. 171). If collapsed post-measurement states are mere 'predictive instrumentalities' to predict the outcomes of further measurements, they are *not* agent-independent descriptions of objective properties of the systems they are assigned to. Since post-measurement situations belong to those cases where we can reasonably hope to have the most complete knowledge and control of the systems under consideration, there is only a very small step from this idea to a fully fledged non-ontic view of quantum states that conceives of *all* quantum states as nondescriptive. This already indicates that a non-ontic reading of quantum states may go very well with the premises of unpristine interpretations, as recommended by Ruetsche on independent grounds.

There are further reasons why unpristine interpretations go well with non-ontic accounts of quantum states: unpristine interpretations hold that what quantum states are to be considered possible for a quantum system depends on the context of state assignment, the 'contingent application', as Ruetsche calls it. Thus, in an unpristine interpretation the range of quantum states to be considered for potential assignment to a quantum system is constrained in a way that depends on the situation of the agent assigning the state. It is natural to extend this conception to the idea that the ultimate narrowing down of the quantum state to be assigned – from the merely 'possible' ones to the 'actual' one – remains sensitive to the situation of the assigning agent, including perhaps her epistemic condition. This, again, is exactly the position taken by non-ontic accounts of quantum states such as the Rule Perspective.

If, in addition to subscribing to a non-ontic account of quantum states, one accepts inferentialism about semantic content as it underlies Healey's pragmatist interpretation (which I do not; see Section 6.1), there is a further parallel with Ruetsche's characterisation of unpristine interpretations, according to which the 'contingent application of theories does not *merely select* among some preconfigured set of their contents, but *genuinely alters* their contents' (Ruetsche [2011], p. 147). According to Healey's inferentialist account of the semantic content of NQMCs, 'the content of S_{or} [some NQMC] is a function of the material inferences that connect it to other claims and other actions by a claimant or others in the same linguistic community' (Healey [2012a], p. 746). But this means that not only the content of the NQMC itself but also that of the theory as a whole is not at all 'preconfigured', to use Ruetsche's word, but rather depends crucially on the material inferences that are legitimate in the context where the theory is put to use. It seems natural to conclude that, for those who feel drawn to Healey's view, the semantic content of language used in quantum theory cannot be specified independently of its 'contingent application' – in complete accordance with Ruetsche's conception of an 'unpristine interpretation' of quantum theory.

To conclude, *if* there are independent good reasons for adopting an unpristine interpretation of the quantum theories of systems with infinitely many degrees of freedom, they increase the attractiveness of adopting a non-ontic reading of quantum states. The concluding section of this chapter presents and discusses the independent reasons for adopting an unpristine interpretation that Ruetsche provides.

11.5 The case for an unpristine interpretation

Ruetsche's main argument for unpristine interpretations, referred to by her as the 'Coalesced Structures Argument', is centred around the claim that no pristine interpretation can account for what she calls 'phase structure' – the coexistence of macroscopically distinct equilibrium phases at certain values of parameters such as coupling constants of a theory or temperature. The phenomenon of phase structure is ubiquitous in the macroscopic objects and substances which surround us in our everyday world. Situations where substances undergo transitions from one phase of matter to another and phase structure manifests itself happen all the time around us. Examples of phase structure include the coexistence of paramagnetic and ferromagnetic phases in iron at 1033 K, that of water and ice at 273 K, and that of water and water vapour at 373 K.

In order for genuine phase structure to occur, the number of degrees of freedom of a physical system must be infinite. If this condition does not hold, all expectation values of observables pertaining to the complete system are analytic (i.e. infinitely-often continuously differentiable) as functions of the parameters characterising the conditions in which the system is placed. Example of such parameters include the system's temperature, the pressure to which it is subjected, or the applied magnetic field. Since the thermodynamic properties of the system can be identified with some of these expectation values, this means that, in the absence of infinitely many degrees of freedom, the thermodynamic properties themselves vary only analytically (i.e. 'smoothly').

However, since phase transitions correspond to non-analyticities in the expectation values that correspond to the thermodynamic properties, there can be no phase transitions – and a fortiori no phase structure – in the presence of only finitely many degrees of freedom. Thus, in order to account for phase structure in the rigorous sense of the word one must take the *thermodynamic limit* of quantum statistical mechanics, where the system size is taken to infinity while the ratio between particle (or component) number N and volume V is kept constant.

In the algebraic approach to quantum theories, the observables in terms of which one defines the identities of the distinct phases are called 'macroobservables'. Macroobservables are characterised by the fact that their value is constant over the whole system in a given phase. An example of a discontinuity signalling a phase transition is the jump in the density of water when turning from the liquid into the gaseous state or the jump in the magnetic susceptibility of iron when turned from the paramagnetic into the ferromagnetic phase.

Troubles for Hilbert Space Conservatism start with the observation that states corresponding to distinct ('pure') phases have different folia, that is, that they cannot be written as density matrices on a single joint Hilbert space $\mathcal{H}$. Assuming that different phases are physically possible for one and the same physical system therefore means to regard more states as physically possible than can be represented as density matrices on a common Hilbert space. This implies that Hilbert Space Conservatism cannot account for phase structure by not acknowledging a sufficient amount of quantum states as physically possible.

Algebraic Imperialism faces a different challenge when attempting to account for phase structure. While acknowledging all the necessary states corresponding to distinct phases, the position has no room to interpret the 'macroobservables' in terms of which of the phases are defined as observables proper. The reason for this is that the defining properties of macroobservables rule out that they be elements of the algebra $\mathcal{A}$, so that macroobservables simply are not observables in the Algebraic Imperialist's sense. Algebraic Imperialism therefore lacks a criterion to distinguish, and therefore to define in the first place, the different phases the existence (and putative coexistence) of which is to be explained. Universalism, as Ruetsche demonstrates, encounters very similar problems. An additional difficulty that concerns both Algebraic Imperialism and Universalism – the 'W^* argument', as Ruetsche calls it – is that not all of the states which they acknowledge as physically possible have a well-defined time-evolution on their GNS representation. To consider states admitting no time-evolution as physically possible seems inherently problematic.

An immediate objection to this line of thought is that the real physical systems, to which the conceptual apparatus of quantum statistical mechanics is applied in practice, have only finitely many degrees of freedom. To describe them in terms of infinitely many degrees of freedom may appear to be an idealisation on which no substantive interpretive claims should be based.[91] Ruetsche herself tends towards the alternative view that interpretations on the assumption of the thermodynamic limit *are* legitimate in that taking this limit is vital to the explanatory success of quantum statistical mechanics.[92] She acknowledges, however, that the 'idealization complaint' (Ruetsche [2011], p. 330) against her interpretive claims must be taken seriously and I follow her in doing so.

To defend the Coalesced Structures Argument against the 'idealization complaint', one may turn to the phase structure exhibited by the theories that are formulated in the language of conventional (Lagrangian) quantum field theory, in particular those that underlie the predictions of modern particle physics. Since these theories are set in a space-time continuum, the number of degrees of freedom is infinite in these theories even in a finite space-time volume.

Phase structure is often approached in terms of the important notion of *spontaneous symmetry breaking* (SSB).[93] Intuitively, a state exhibits SSB just in case the symmetries of the underlying laws of motion are *radically* absent from it: the Lagrangian (or Hamiltonian) of the theory is invariant under certain transformation, and the symmetry-breaking state cannot be changed to one which exhibits this symmetry by means of 'physically realizable operations' (Strocchi [2008], p. 10). SSB is closely related to phase structure in that the distinction between states of broken and unbroken symmetry usually lines up with a distinction between different phases. In conventional quantum field theory (QFT), SSB is widely believed to play a fundamental role in the generation of particle masses in form of the 'Higgs mechanism', which is often associated with the notion of a *spontaneously broken local gauge symmetry*.[94]

Ruetsche acknowledges two potential weaknesses of her strategy for arguing against *pristine interpretations* by appeal to the Coalesced Structures Argument and conventional QFT. The first potential weakness is that, at present at least, we do not have any rigorous formulation of the most successful theories of conventional QFT in the algebraic framework. The theories combined in the standard model of elementary particle physics are nowadays widely regarded as *effective field theories*, i.e. theories whose range of validity does not extend to arbitrarily high energies and arbitrarily small distances. They are commonly studied in the framework of lattice gauge theory, using, once again, the methods of quantum statistical mechanics. Therefore, the question of whether physicists' QFT supports the case against pristine interpretations is at least in part a matter of speculation about future physics, though arguably one where an optimistic stance seems not unreasonable: it would be very astonishing indeed if the phase structure computed based on discrete approximations to the continuum did vanish once the features of the actual continuum (should it actually exist as such) are properly taken into account.

The second potential weakness Ruetsche sees in her argument is that the alleged explanatory role of SSB, and hence the status of phase structure in conventional QFT, remains unclear inasmuch as the status of SSB in the generation of particle masses through the Higgs mechanism continues to be a matter of debate. She takes this obstacle much too seriously, however, due to a misunderstanding about gauge symmetry breaking and the Higgs mechanism to which she is led by identifying a specific choice of gauge (comparable in its significance to the choice of a coordinate system), namely, the *unitary* gauge, with 'the state of broken symmetry suiting the Higgs mechanism' (Ruetsche [2011], p. 331). This is misleading, first, in that a gauge is not a 'state' at all, in particular not one exhibiting broken symmetry, and, second, in that making the specific choice of the *unitary* gauge specifically means eliminating the gauge symmetry completely at an 'explicit' level, so that SSB cannot occur at all.[95] Other choices of gauge fixing do not eliminate the gauge symmetry completely and allow that a remnant global *subsymmetry* of the local gauge symmetry can still be spontaneously broken. However, the relation between these instances of SSB and phase structure is very complicated in that the distinction between broken and unbroken remnant global gauge symmetry is not in general associated with a distinction between different phases.[96]

As far as *local* gauge symmetries themselves are concerned, it should be noted that our current frameworks of quantising gauge theories do not support the notion of a spontaneously broken local gauge symmetry at all. This statement can be made rigorous in the context of lattice gauge theory, where it is known as 'Elitzur's theorem' (Elitzur [1975]). However, regardless of what one actually sees as the role of SSB in the Higgs mechanism or in particle physics more generally, it is uncontroversial that our most successful theories of conventional QFT exhibit a very rich phase structure.[97] To conclude, there seems to be strong ground for optimism that these theories, once cast in a rigorous formulation using the algebraic framework, will after all exhibit, in the words of Ruetsche, 'a structure sustaining the Coalesced Structures Argument' (Ruetsche [2011], p. 335).

And the conclusion of this 'Coalesced Structures Argument', as argued in the previous section, provides further motivation for adopting a non-ontic account of quantum states such as the Rule Perspective.

12
Quantum Theory and 'Reality'

The main ambition of this work has been to explore the prospects for dissolving the foundational problems of quantum theory by adopting the epistemic conception of quantum states and reflecting on how it should be spelled out in detail. This differs (at least in emphasis, if not more profoundly) from the main ambition of those – 'realistically minded' – philosophers whose declared aim is to determine what reality is like if it is as quantum theory says. If the Rule Perspective is correct, quantum theory does not really *say* anything about reality at all: features of reality are described not by quantum states, but by NQMCs, and while quantum theory helps us ascribe probabilities to the NQMCs, it does not typically make any statements as to which NQMCs are true and which are false. However, even though quantum theory does not in that sense *say* anything about reality, we can learn a lot about reality by reflecting on what features it must have for quantum theory's undisputed empirical success to be possible.

The present, final, chapter of this work offers such reflections. In the first section it explains why adopting the Rule Perspective does not mean falling back on an uninteresting instrumentalism. In the second section it argues further that the Rule Perspective, perhaps somewhat unexpectedly, may provide a hint as to why and how the ambitiously 'realist' project of attributing sharp values to all observables of all quantum systems at all times (and spaces) may be realisable after all – even though this project admittedly goes beyond the boundaries of the therapeutic approach pursued here, which explores whether quantum theory may be 'fine as it stands'. The third and final section completes the book with a brief summary and concluding remark.

12.1   Instrumentalism?

Some worries as to whether the Rule Perspective might be no more than unilluminating instrumentalism about quantum theory were already answered, at least implicitly, when I argued in Chapter 9 that the Rule Perspective is able to account for quantum theory's explanatory force no less than for its predictive force. Further progress towards answering the challenge of instrumentalism can be made along the lines suggested by Timpson, who responds to parallel complaints against quantum Bayesianism:

> Quantum mechanics may not be a descriptive theory, we may grant, but it is a significant feature that we have been driven to a theory with just this characteristic (and unusual) form in our attempts to deal with and systematize the world. The structure of that theory is not arbitrary; it has been forced on us. Thus by studying in close detail the structure and internal functioning of this (largely) non-descriptive theory we have been driven to, and by comparing and contrasting with other theoretical structures, we may ultimately be able to gain *indirect* insight into the fundamental features of the world that were eluding us on any direct approach. (Timpson [2008], pp. 582f.)

What are the 'fundamental features of the world' quantum theory reveals to us while they were 'eluding us on any direct approach'? An answer to this question that seems to be in the spirit of Timpson's words is given by Jeffrey Bub, whose position invites a similar charge of instrumentalism to quantum Bayesianism in that it also interprets quantum probabilities as subjective degrees of belief. Bub proposes that the non-Boolean nature of the lattice of projection operators on the Hilbert spaces employed in quantum mechanics should be seen as corresponding to 'an objective feature of the world, the fact that events are structured in a non-Boolean way' (Bub [2007], p. 252). This suggestion is vague, however, and, in addition, it is unclear whether a proponent of the Rule Perspective should agree with it. For the Rule Perspective holds that quantum probabilities are defined only over those NQMCs that are 'licensed' (as Healey puts it) by quantum theory if effects of environment-induced decoherence are properly taken into account, and these NQMCs correspond to a *Boolean* sublattice of the complete (non-Boolean) lattice of projection operators.

A more promising approach to specifying in which sense quantum theory may give us '*indirect* insight into the fundamental features of the world' (Timpson's words) is by considering in virtue of which features of reality the objective probabilities of events can be those that quantum theory gives us. If we adhere to the Principal Principle, as I think we should, this is the question of in virtue of which features of reality quantum probabilities are the rational credences for the agents using quantum theory.

Lewis provides his account of the features of reality in virtue of which objective probabilities are what they are as an element of his *best-system* analysis of laws of nature. Laws of nature, on a best-system analysis, are those statements that provide the best true systematic account of the world, superior to all others by the standards of 'simplicity, strength, and fit' (Lewis [1994], p. 480).

The best-system analysis of laws of nature has a number of attractive features; perhaps its most evident virtue is that it goes so well with how we actually try to determine the laws of nature, namely, by emphasising regularities among the data we have accumulated and then using criteria such as simplicity, strength, and fit to infer the laws from them. It thus fits well with scientific practice, while avoiding metaphysical speculation that is alien to that practice. Of course, like all other types of approach in philosophy, the best-system analysis of laws faces serious difficulties, but, as convincingly argued by (Cohen and Callender [2009]), the prospects for answering those that are widely regarded as the most serious ones are quite good.

Best-system accounts of objective *probability* form a special class of best-system accounts of laws of nature. What makes them probabilistic is that some of the quantities they involve impose constraints on the rational credences of appropriately situated agents just as objective probabilities need to by the Principal Principle. As highlighted by (Lewis [1994], pp. 481f.), according to his approach the probabilistic laws obtain in virtue of the *relative frequencies* and the *symmetries* in the overall pattern of events across space-time. Conversely – and this is the crucial point to note in the present context – we can draw some conclusions about the symmetries and (approximate) relative frequencies in the pattern of events once we know how to make probability ascription in the light of the best theory of reality that we have, e.g. quantum theory.

Lewis himself offers his best-system account of probabilistic laws in combination with a refined version of the Principal Principle (Eq. (NP) in (Lewis [1994], p. 487)), which addresses a specific technical weakness of the Principal Principle in its original formulation that can safely be ignored for our present purposes. His account is further developed in a seminal contribution by (Hoefer [2007]), who, like Lewis, proposes his account of objective probabilities in the framework of the thesis called 'Humean Supervenience'. According to this thesis 'the whole truth about a world like ours supervenes on the spatio-temporal distribution of local qualities' (Lewis [1994], p. 473), and this complete spatio-temporal distribution of 'local qualities' is referred to as the 'Humean mosaic'. Hoefer points out that objective probabilities apply to the possible 'local qualities' of the pattern only in virtue of their being elements of a 'setup' or 'reference class':

> Some linking of objective probability to a setup, or a reference class, is clearly needed. Just as a Humean about laws sees ordinary laws as, in the first instance, patterns or regularities – in the [Humean] mosaic, whenever F, then G – so the Humean about chances sees them as patterns or regularities in the mosaic also, albeit as patterns of a different and more complicated kind: whenever S, $Pr(A) = x$. (Hoefer [2007], p. 564)

The conception of a 'Humean mosaic' that is composed of 'local qualities' is deeply metaphysical and, as such, operates on an entirely different level than the Rule Perspective. Nevertheless, starting from the Rule Perspective, the ideas of Hoefer (and Lewis) can help to specify in which way quantum theory may give us 'indirect insight' (to use Timpson's words) into the nature of reality. The essential idea here is that, in order for quantum theory to have the empirical success that it has, the patterns and regularities among the values of observables must be such that, for appropriately situated agents, precisely those credences are the rational ones which quantum theory specifies for them in the form of the Born Rule probabilities.

Hoefer's point that any probability ascription must be linked 'to a setup, or a reference class' is recovered in the Rule Perspective in terms of Healey's insight that Born Rule probabilities apply only to those NQMCs that are 'licensed' for the agents who ascribe them. Here, to repeat, for some NQMC to be 'licensed', the (reduced) density matrix assigned to the system must be at least approximately diagonal in a preferred way in an eigenbasis to the observable referred to in the NQMC if effects from environment-induced decoherence are taken into account. Circumstances in which NQMCs are 'licensed' are candidate 'setups' for objective probability ascriptions in Hoefer's sense. In quantum theory his formal condition 'whenever S, $Pr(A) = x$' becomes 'whenever conditions are such that ρ must be assigned by some (actual or hypothetical) agent and the NQMC "The value of (*A*) lies in Δ" is licensed, the Born Rule applies in that the probability of the value of *A* lying in Δ is given by $\mathrm{Tr}(\rho\,\Pi_\Delta^A)$'.

Quantum theory, according to this account, does not directly tell us anything about the distribution of the values of observables. Nevertheless, in the spirit of the passage by Timpson quoted above, we can learn something about this distribution in an *indirect* way: that the relative frequencies and the symmetries in it are such that the rational credences for agents are given by the Born Rule probabilities whenever the setup conditions are fulfilled. The smaller the range Δ for which NQMCs of the form 'The value of *A* lies in Δ' are 'licensed', the more we learn from quantum theory about the features of the distribution. And we do learn a lot: the ranges Δ attributed to the values of observables in many NQMCs that are 'licensed' in practice are sufficiently narrow for quantum theory to have the tremendous empirical success that it manifestly has.

12.2 Sharp values for all observables?

As explained, the Rule Perspective follows Healey in his insight that Born Rule probabilities apply only to those observables for which the reduced density matrix assigned to a system becomes (at least approximately) diagonal in a preferred basis that is typically selected by environment-induced decoherence effects, if they exist and are taken into account. The NQMCs that attribute sharp values to these observables are the ones Healey characterises as those that are 'licensed' in the given context of assignment.

Up to now, the present work has relied on a minimalist reading of what is involved in the 'licensing' of a NQMC of the form 'The value of *A* lies in Δ'. The minimalist reading leaves it open whether even for a NQMC that is *not* licensed for any (actual or hypothetical) physically situated agent the observable *A* *has* some determinate value, whether inside or outside the range Δ. Possible alternatives to this minimalist reading are: first, the austerely 'realist' view according which *all* observables have sharp values in all circumstances – even those that are not licensed for any physically situated agent; second, to *deny* that in those cases where there is no such agent (whether actual or hypothetical) *A* has any determinate value; and, third, to claim that the very question of whether *A* has some determinate value is misleading, perhaps meaningless.

Here I wish to explore the prospects for the austerely 'realist' view according to which all observables have sharp values in all circumstances – even those that are not licensed for any actual or hypothetical agent. The idea of a complete assignment of sharp values to all observables evidently goes beyond the boundaries of the therapeutic approach to quantum theory pursued here so far according to which the theory needs no further theoretical additions. Nevertheless, we may ask whether all the *other* core tenets of the Rule Perspective besides its therapeutic motivation – its account of quantum states, quantum probabilities, the Born Rule, etc. – are compatible with this idea. Arguably yes, as I shall argue in what follows, though the issue remains complicated and somewhat speculative. If I am right about this, there is no real conflict at all between the Rule Perspective and what is perhaps the most ambitiously realist desideratum one might have when interpreting quantum theory.

To begin with, the idea of a complete assignment of sharp values to all observables of a quantum system may seem hopelessly naive in view of the famous no-go theorems due to Bell, Gleason, Kochen, Specker, and many others. However, on closer inspection none of these theorems seems to exclude that such an assignment might be made in an uncontrived and elegant way in conformity with quantum theory as seen from the standpoint of the Rule Perspective (minus the latter's underlying therapeutic motivation).

To see this, let us first have a look at the Kochen–Specker theorem (Appendix B). It rules out that assignment of sharp values to observables can be made in such a way that the functional relations among the possessed values of observables are the same as those among the Hilbert space operators corresponding to the observables, even if only mutually compatible observables are considered. This is often summarised by saying that the Kochen–Specker theorem requires assignments of values to observables to be *contextual* in that they must take into account the physical (for example, measurement) context of the system at issue in a non-trivial way (See Appendix B for the details of how the assumption of non-contextuality enters the derivation of the Kochen–Specker theorem).

According to the Rule Perspective, the Born Rule does not apply to those NQMCs that are *not* 'licensed' for a quantum system and, so, does not specify any probabilities for them. For those NQMCs which, in contrast, are 'licensed', the Born Rule tells us that the only possible values of the observables they refer to are the eigenvalues of the associated Hilbert space operators. However, if we probe the hypothesis that the observables referred to in those NQMCs that are *not* 'licensed' have sharp values, quantum theory in general and the Born Rule in particular are irrelevant, and there is no reason at all to suppose that the possible values of these observables are among the eigenvalues of the associated Hilbert space operators. Quantum theory, as conceived of by the Rule Perspective, does not specify any probabilities for the values of these observables (i.e. those referred to in NQMCs that are not 'licensed'), so we should not assume that the eigenvalues of the associated Hilbert space operators are *in general* the only possible values of observables.

Let us return to the Kochen–Specker theorem. It is an immediate consequence of the 'functional composition rule' FUNC (see Appendix B), which plays an important role in the derivation of the theorem, that the eigenvalues of the operator that the quantum theoretical formalism associates with an observable are in all circumstances the observable's only possible values. In the light of the considerations just presented, we have little reason to believe that this consequence of FUNC holds and, thus, little reason to believe in FUNC itself. Since FUNC can be seen as the mathematical expression of non-contextuality, non-contextuality in (hypothetical) complete assignments of sharp values to observables is to be expected from the point of view of the Rule Perspective. Thus, the Kochen–Specker theorem poses no threat to the idea that sharp values might be assigned to all observables in a natural and coherent way and, at the same time, in agreement with quantum theory in the light of the Rule Perspective.

Similarly, Bell's theorem poses no threat to this idea. This result concerns *any* theory which reproduces the quantum predictions about correlations, and it tells us that any such theory must violate Bell's formulation of local causality $Pr(A|\lambda) = Pr(A|\lambda, B)$, where A and B occur at space-like distance from each other and λ fully specifies A's backward light cone (or a subregion of the latter that screens off A from B's backward light cone).[98] As argued in detail in Chapter 10, Bell's formulation does not adequately spell out the idea that the probability of an event depends only on what occurs in its backward light cone and is in that sense not appropriately characterised as a criterion of 'local causality' at all. There is no reason to suppose that (hypothetical) complete assignments of sharp values to observables would inevitably lead to a violation of local causality, properly construed, where quantum theory does not. Given these insights, there is no reason to suppose that assignments of sharp values to observables would lead to a clash with the principles of relativity theory, where quantum theory by itself does not. Bell's theorem, to sum up, gives us no reason to suppose that assignments of sharp values to all observables of a quantum system are impossible or that they cannot be made in a manner compatible with special relativity.

Finally, let us consider whether the recently proved PBR theorem (Pusey et al. [2012]; for a brief introduction see Appendix C) points to insurmountable obstacles for assignments of sharp values in accordance with the Rule Perspective. One of the assumptions on which this theorem rests says that for any quantum system its complete physical state λ can be written down, which – we may suppose – includes a complete collection of the values of its observables. As the proof of the theorem shows, given two apparently very plausible further assumptions (both to be mentioned in a moment), the 'prepared' quantum state ψ must be part of the complete state λ if the predictions of quantum theory are to be recovered. In other words, the preparation procedures associated with any two distinct quantum states ψ_1, ψ_2 must give rise to physically distinct complete states λ_1, λ_2. Thus, given the assumptions taken together, the complete physical state λ uniquely determines the quantum state ψ that has been prepared, which means that the latter can apparently be seen as a part and aspect of the former (or identical to it). And this means that the quantum state is itself physical and hence 'non-epistemic', contrary to what the Rule Perspective assumes. So, if the assumptions of the theorem hold, it follows that the Rule Perspective cannot be combined with the hypothesis that all observables have sharp values without undermining its own presupposition that quantum states are 'epistemic'. Can this diagnosis be avoided?

Arguably yes, since the further two assumptions used in the proof of the PBR theorem need not hold for complete assignments of sharp values to all observables, however natural and appealing they may appear at first glance. The first further assumption, 'ψ independence' (so named by (Schlosshauer and Fine [2012])), says that measurement outcomes depend only on the physical state λ and are not correlated with the preparation method used to obtain that state. This is an assumption that we need not take for granted when considering complete assignments of sharp values to observables. Perhaps correlations in the spatio-temporal pattern of values of observables are such that the physical state does not 'screen off' the preparation method from the measurement outcomes. This may appear odd, because physical systems we are familiar with do not seem to exhibit this type of 'memory' concerning the preparation method used. But perhaps this oddity merely reflects an implicit bias on our part, shaped by our everyday experiences and best not presupposed when assessing the viability of sharp value assignments to observables at (what one may call) the level that is unconstrained by the Born Rule probabilities.

The second further assumption that is used in the proof of the PBR theorem, 'preparation independence', is an assumption of statistical independence for the physical states λ_i of quantum systems for which, individually, preparation procedures corresponding to the quantum states $\psi_1, \psi_2, \psi_3, \ldots$ are chosen (See Appendix C for a few more details).

Independence assumptions of this kind appear natural and legitimate in 'everyday' scientific contexts, but they remain speculative when it comes to assignments of sharp values to observables over and above those constrained by the Born Rule probabilities.[99] Thus, while the PBR theorem highlights interesting and important constraints on assumptions about physical states of quantum systems (including complete assignments of sharp values to all observables), it does not by any means exclude the possibility of combining the Rule Perspective with a complete assignment of sharp values to all observables.

There are further interesting and important no-go theorems about physical states of quantum systems that concern hypothetical assignments of sharp values to all observables; see, for instance, the results due to (Leggett [2003]), (Hardy [2004]), and (Conway and Kochen [2006]). However, the assumptions used to derive these theorems are no less contestable than those that underlie the Kochen–Specker theorem and the PBR theorem.[100]

To conclude, there is at present no reason to doubt that appealing and uncontrived assignments of sharp values to all observables are possible and compatible with quantum theory as conceived of from the standpoint of the Rule Perspective. Lacking any further details, the conception of such an assignment is at present somewhat speculative, but this should arguably not prevent us from entertaining it.

Two scenarios of in which way assignments of sharp values to all observables may one day turn out possible come to mind. In the first, physicists ultimately arrive at a theory that, first, is more fundamental than quantum theory, that, second, reproduces the quantum theoretical predictions over the whole range of phenomena where quantum theory is successful, and, third, delivers – as a byproduct, so to speak – complete assignments of sharp values to all quantities that it countenances as 'physical', including many, if not all, observables of all quantum systems. We cannot exclude that such a theory will one day be found by physicists. Should it happen, all the foundational problems of quantum theory would have to be seen in the light that is thrown on them by the more fundamental theory.

In the second, alternative, scenario of how complete assignments of sharp values to observables may turn out possible, there will be no systematically privileged more fundamental *theory* than quantum theory that produces such value assignments. Such assignments, even if they can be shown to be feasible in a natural and uncontrived way, may remain underdetermined by the standard criteria of theory choice such as empirical adequacy, elegance, and simplicity. If this scenario becomes real, the idea of a complete description of the world in terms of a complete specification of the values of all observables remains a live and coherent one. Nevertheless, unlike quantum theory itself, the details of that description might be forever left underdetermined by the data.

Independently of which of these scenarios seems more plausible and which, if any, is ultimately going to hold true, let me conclude this section by affirming once again that the prospects for assigning sharp values to all observables are not at all unpromising when considered in the light of the Rule Perspective. The task of actually specifying such assignments remains open for the future as a challenge.

12.3 Conclusion

The project of approaching the foundations of quantum theory in a therapeutic spirit has brought us a long way. After giving a sketch of the most elementary features of the mathematical formalism of quantum theory I outlined the historical and methodological background of the therapeutic approach to philosophical problems that is associated, in particular, with the later Wittgenstein. Next, I explained why, when trying to apply this approach to the foundations of quantum theory, an obvious first step is to embrace the epistemic conception of quantum states. This, however, gave rise to the challenge of specifying how that conception should best be spelled out in detail. In response to this challenge I presented the Rule Perspective as the, to my mind, most promising epistemic account of quantum states due to how it reconciles the concept of a state assignment being performed *correctly* with the rejection of the notion of the quantum state a quantum system 'is in'. The core ideas of the Rule Perspective are, first, to conceive of the rules governing state assignment as constitutive rules in Searle's sense and, second, to conceive of the range of applicability of the Born Rule as limited to the values of those observables with respect to which the density matrix assigned to the system is at least approximately diagonal, typically as a consequence of environment-induced decoherence.

Having argued that the Rule Perspective is not merely a restatement of the notoriously elusive Copenhagen interpretation in new clothing, I proceeded to defend it against various objections, in particular against that of illegitimately relying on primitive anthropocentric notions and against that of being unable to account for quantum theory's incontestable explanatory force. Next, I considered the alleged tension between quantum theory and relativity theory due to 'quantum non-locality', reaching the conclusion that, for all we know, these two 'pillars of contemporary theory' (Bell [2004], p. 172) are perfectly compatible. Moreover, as I argued, we have no reason not to regard quantum theory as perfectly 'locally causal' in that quantum probabilities, properly viewed, depend only on what occurs in the backward light cone. Considerations about algebraic quantum field theory, partly based on Laura Ruetsche's case in favour of what she calls 'unpristine interpretations' of quantum theories, were adduced as further support for a non-ontic view of quantum states. In the present, final, chapter, I explored whether quantum theory, as conceived of from the point of view of the Rule Perspective, is compatible with the ambitions of those who look for a more 'robustly realist' view of the theory and came to the conclusion that it is.

In summary, I conclude that the route towards dissolving the quantum paradoxes by adopting an epistemic account of quantum states is open and promising. The role of the most characteristically 'quantum' vocabulary – quantum states – may not be that of describing physical reality, but, if so, this does not make the theory any less interesting. Quite the contrary: if something close to the Rule Perspective is correct, not only *what* we learn from quantum theory about reality is intriguing and fascinating, but also *how* it is that we do so.

Appendices

Appendix A
Sketch of Bell's Theorem

The label 'Bell's theorem' refers to a family of results – originating from the pioneering work of J. S. Bell as collected in (Bell [2004]) – which show that any theory that reproduces the quantum theoretical predictions based on entangled states must be 'non-local' in the sense of violating a criterion Bell refers to as 'local causality'.

Here I present the statement of Bell's theorem and a sketch of its derivation in three steps.[101] First, I briefly recapitulate the motivation and content of Bell's criterion of local causality. Second, I derive the nowadays most commonly used ancestor of an inequality derived by Bell, called the 'Clauser–Horne–Shimony–Holt inequality' or 'Bell–Clauser–Horne–Shimony–Holt inequality' (BCHSH inequality), which imposes constraints on the expectation values of products of observables in theories that conform to Bell's criterion. Third, I give an outline of in which way quantum theory violates the BCHSH inequality.

Deriving the BCHSH inequality

Consider a situation where measurements of quantum systems are performed by two spatially separated agents Alice and Bob (who may be at space-like separation from each other; see Figure 2, an adaptation from Figure 1 in Section 10.2) with measured observables (apparatus settings) $A, A', \ldots$ and $B, B', \ldots$ and outcomes $a, a', \ldots$ and $b, b', \ldots$ (Note that this convention of using uppercase letters for observables and lowercase letters for outcomes, introduced in Chapter 2, is opposed to the common one in the literature on Bell's theorem). As explained in Section 10.2.2, the essential idea behind Bell's probabilistic criterion of local causality is that in a locally causal theory the probability of the outcome a in region 1 depends only on what occurs in the backward light cone of region 1. Bell implements this idea by requiring that, given a complete description of what occurs in a region of the backward light cone of region 1 that 'completely shields off' (Bell [2004], p. 240) region 1 from the backward light cone of the space-like separated region 2 (such

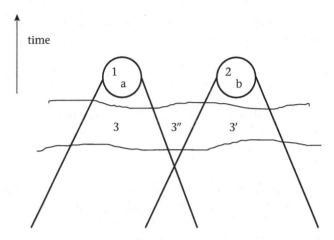

time

Figure 1.1 In a theory that is locally causal in Bell's sense, giving a complete specification of what occurs in regions 3, 3', and 3" makes information about what happens in region 2 irrelevant for the probabilities in region 1

as region 3 in Figure A.1), the result of the measurement in region 2 is irrelevant to the probabilities of possible outcomes in region 1.

If we assume that what occurs in regions 3, 3', and 3", taken together, has been given a complete[102] description λ, the assumption with which the derivation of the BCHSH inequality starts is

$$P_{A,B}(a|b,\lambda) = P_A(a|\lambda),\tag{A.1}$$

$$P_{B,A}(b|a,\lambda) = P_B(b|\lambda),\tag{A.2}$$

where the indices A and B indicate the observables that are measured. (Since the description λ concerns even regions outside the backward light cone of region 1, the result that quantum theory violates Eqs. (A.1) and (A.2) is, strictly speaking, stronger than that it violates local causality according to how Bell characterises this notion. Nevertheless, the violation of local causality in any theory that reproduces quantum theory can be inferred directly from the violation of Eqs. (A.1) and (A.2), which is why I elsewhere refer to these equations, somewhat loosely, as encoding Bell's criterion of local causality.)

Eqs. (A.1) and (A.2) jointly imply the factorisation condition:

$$P_{A,B}(a,b|\lambda) = P_A(a|\lambda)P_B(b|\lambda).\tag{A.3}$$

At this point, it is useful to define shorthands for the expectation values of possible outcomes of measurements of a and b, namely,

$$E_{A,B}(\lambda) = \sum_a \sum_b ab P_{A,B}(a,b|\lambda). \tag{A.4}$$

Using the factorisation condition Eq. (A.3) they can be rewritten as

$$E_{A,B}(\lambda) = \sum_a a P_A(a|\lambda) \sum_b b P_B(b|\lambda). \tag{A.5}$$

Now, in terms of these expressions one can derive the BCHSH inequality. Consider apparatus settings A and B so that possible measured values are all in the closed interval $[-1,1]$. It is then an elementary mathematical fact that for any real numbers a, b, a', b' within [-1,1] the inequality $|ab + ab' + a'b - a'b'| \leq 2$ holds. Due to the factorisation condition Eq. (A.3) this gives the following (preliminary) inequality:

$$|E_{A,B}(\lambda) + E_{A,B'}(\lambda) + E_{A',B}(\lambda) - E_{A',B'}(\lambda)| \leq 2. \tag{A.6}$$

If we now assume that *statistical independence* holds in the sense that the probability of the state λ is unaffected by what observables A and B are measured at the systems, Eq. (A.6) can straightforwardly be translated into a statement in terms of probability distributions ρ over the space Λ of all possible complete descriptions $\lambda \in \Lambda$ of the backward light cone of region 1. This is achieved by integrating over Λ with an integration measure $d\rho$ defined by $p_\rho(a,b|A,B) \equiv \int_{\lambda \in \Lambda'} p_\lambda(a,b|A,B) d\rho$. Defining

$$E_{A,B}(\rho) \equiv \int_{\lambda \in \Lambda} E_{A,B}(\lambda) d\rho, \tag{A.7}$$

this results in the BCHSH inequality, which says

$$|E_{A,B}(\rho) + E_{A,B'}(\rho) + E_{A',B}(\rho) - E_{A',B'}(\rho)| \leq 2. \tag{A.8}$$

Quantum mechanical violation of the BCHSH inequality

Quantum theoretical predictions are naturally regarded as in conflict with the BCHSH inequality and, thus, as violating Bell's criterion of local causality Eqs. (A.1), (A.2). Consider a two-component system with subsystem Hilbert spaces $\mathbb{C}^2$ and observables A, A', B, B' given by

$$A = S_x \otimes I, \tag{A.9}$$

$$A' = S_z \otimes I, \tag{A.10}$$

$$B = -\frac{1}{\sqrt{2}} I \otimes (S_z + S_x), \tag{A.11}$$

$$B' = \frac{1}{\sqrt{2}} I \otimes (S_z - S_x). \tag{A.12}$$

Consider now the following entangled state of the two-part system:

$$|\phi\rangle = \frac{1}{\sqrt{2}} (|+x\rangle_1 |-x\rangle_2 - |-x\rangle_1 |+x\rangle_2). \tag{A.13}$$

The following expectation values can be derived from that state:

$$\langle AB \rangle_\phi = \langle A'B \rangle_\phi = \langle A'B' \rangle_\phi = \frac{1}{\sqrt{2}}, \tag{A.14}$$

$$\langle AB' \rangle_\phi = -\frac{1}{\sqrt{2}}, \tag{A.15}$$

which implies

$$\langle AB \rangle_\phi + \langle A'B \rangle_\phi + \langle A'B' \rangle_\phi - \langle AB' \rangle_\phi = 2\sqrt{2} > 2. \tag{A.16}$$

If one interprets the quantum theoretical expectation values in Eq. (A.16) as expectation values derived from probability distributions ρ, performed over states λ, one will find that this is in contradiction with the BCHSH inequality Eq. (A.8).

Quantum theoretical predictions about correlations such as those expressed in the expectation values occurring in Eq. (A.16) are well-confirmed experimentally, most famously in the experiments directed by Alain Aspect; see (Aspect et al. [1982]). Thus, by this line of reasoning, quantum theory as well as any theory that reproduces its predictions must inevitably violate Bell's formulation of 'local causality' as expressed in Eqs. (A.1) and (A.2).

Appendix B
The Kochen–Specker Theorem in a Nutshell

The Kochen–Specker theorem (Kochen and Specker [1967]) highlights significant constraints on the assignment of sharp values to observables. As the proof of the theorem shows, for sufficiently large collections of observables any such assignment of sharp values must necessarily be *contextual*. Contextuality means, roughly speaking, that if some observable has a sharp value, that value depends in some way on the (e.g. measurement) context in which the system is placed.

Statement of the theorem

Here I first state the Kochen–Specker theorem and name the assumptions which are needed to derive it. Further below I briefly discuss the main motivation for these assumptions.

Let $\mathcal{H}$ be a Hilbert space of dimension $n \geq 3$. Then there is a set M of k observables (associated with linear operators on $\mathcal{H}$) with some finite natural number k which depends on n such that the members of M cannot all have sharp values in the range of real numbers, provided the assignments of values $v(A)$, $v(B)$, $v(C)$ to any three compatible observables A, B, C are supposed to fulfil the following requirements: is

(a) If $C = A + B$, then

$$v(C) = v(A) + v(B). \tag{B.17}$$

(b) If $C = AB$, then

$$v(C) = v(A) \cdot v(B). \tag{B.18}$$

The original proof of Kochen and Specker (Kochen and Specker [1967]) concerns Hilbert space dimension $n = 3$ and a set of $k = 117$ observables. For dimension 3, the smallest number k of observables for which a proof is known is $k = 33$ (Peres [1991]); for dimensions $n \geq 4$ the smallest number k for which a proof can be run is known to be $k = 18$. (Cabello et al.

[1996]) I will not rehearse any of these proofs here; see (Held [2013]) for a useful overview.

The foundational relevance of the Kochen–Specker result depends on the plausibility and naturalness of the constraints (a) and (b) on the assumed values of compatible observables. Both (a) and (b) can be obtained as instances of a more general principle known as 'FUNC' ('functional composition principle').[103] To be able to formulate this principle, we have to distinguish between observables and the associated Hilbert space operators. In what follows, whenever a Hilbert space operator is meant, this is indicated by a 'hat', as in '$\hat{A}$', to distinguish it from the observable A.

> FUNC: For any self-adjoint linear Hilbert space operator $\hat{A}$ associated with an observable A and any function $f : \mathbb{R} \mapsto \mathbb{R}$ from the reals to the reals, the self-adjoint operator $f(\hat{A})$ is associated with a unique observable $f(A)$, such that for the values $v(A)$ and $v(f(A))$ the following holds:
>
> $$v(f(A)) = f(v(A)). \tag{B.19}$$

The principle FUNC can be derived using three further assumptions (to be discussed further below) together with the following equality (also known as 'STAT FUNC'[104]), which is itself a theorem of quantum theory:

$$Pr[v(f(A))_\rho = a] = Pr[f(v(A))_\rho = a]. \tag{B.20}$$

This equality holds in the quantum mechanical formalism as a consequence of the Born Rule together with the fact that $\text{Tr}(\hat{\Pi}_a f(\hat{A})\rho) = \text{Tr}(f(\hat{\Pi}_a \hat{A})\rho)$, where $\hat{\Pi}_a$ projects on the linear span of eigenvectors of $\hat{A}$ with eigenvalue a.

Why the functional composition rule?

The three additional assumptions which are necessary to derive the FUNC from STAT FUNC are commonly referred to as *value-definiteness*, *value realism*, and *non-contextuality*. In what follows I state and explain these assumptions.

The assumption of *value-definiteness* says that all observables have sharp values. The main interest of results such as the Kochen–Specker theorem lies in highlighting its implications, so rejecting this assumption is not an option in the present context.

The second assumption, called '*value realism*' by (Held [2013]) and referred to as the '*Reality Principle*' by (Redhead [1987], p. 133), concerns the association of Hilbert space self-adjoint linear operators with observables. It says that, if the quantum theoretical algorithm associates some operationally defined number a with some self-adjoint operator $\hat{A}$ by defining a probability p such that $p = Pr(v(\hat{A}) = a)$ according to the Born Rule, then there exists an observable A of which a is a possible value.

The third assumption – *non-contextuality* – states that observables are in one-to-one correspondence with Hilbert space self-adjoint linear operators.

From the assumptions STAT FUNC, value-definiteness, value realism, and non-contextuality the principle FUNC can be derived as follows: let A be an observable of a quantum system, associated with a self-adjoint linear operator $\hat{A}$. By value-definiteness, it has a sharp value $v(A)$. From STAT FUNC we obtain $Pr[v(f(A))_\rho = a] = Pr[f(v(A))_\rho = a]$ for arbitrary values $v(A)$ of that observable, arbitrary numbers a, and arbitrary functions f. Implicitly, for each f, we have thus constructed a new operator $f(\hat{A})$ that is associated with the numbers a and $p = Pr[v(f(\hat{A}))_\rho = a]$. Value realism allows us to conclude that at least one observable $f(A)$ exists which corresponds to this operator and for which $v(f(A)) = f(v(A))$. Non-contextuality guarantees that there exists exactly one such observable $f(A)$. This yields the functional composition principle FUNC.

Chapter 12 of this work considers a motivation for accepting contextuality which follows naturally from the account developed before. For a systematic approach to the prospects and ramifications of accepting contextuality, see (Held [2013], in particular section 5.3) and references therein.

Appendix C
The Pusey–Barrett–Rudolph (PBR) Theorem – A Short Introduction

The Pusey–Barrett–Rudolph (PBR) theorem (Pusey et al. [2012]) is a recent and already famous no-go result, which establishes that certain very plausible-looking assumptions rule out epistemic accounts of quantum states. More specifically, what the theorem says is that, given the assumption that each quantum system has a complete ('ontic') physical state λ – and two further assumptions to be mentioned further below – this physical state λ uniquely determines the pure quantum state ψ that has been prepared. Otherwise, the predictions of quantum theory could not in general be reproduced.

The theorem and its assumptions

The reasoning underlying the PBR theorem associates quantum states ψ with probability distributions $\mu_\psi(\lambda)$ over possible 'ontic' physical states λ. In terms of these probability distributions the question of whether an epistemic account of quantum states is viable takes the form of whether any two distinct distributions $\mu_{\psi_1}(\lambda)$ and $\mu_{\psi_2}(\lambda)$ can have non-vanishing *overlap*, i.e. whether one and the same physical state λ is compatible with more than one quantum state ψ. If this is the case, then, whenever λ is the physical state of the system, there is no fact of the matter as to which quantum state ψ the system is in. The quantum state ψ is not a physical quantity in this case and in this sense 'epistemic'.

The proof of the PBR theorem shows that the quantum theoretical predictions rule out that quantum states are 'epistemic' in this sense, given the assumptions of 'ψ *independence*' (this terminology is due to (Schlosshauer and Fine [2012])) and '*preparation independence*' (PBR's own term). The assumption of ψ *independence* is quite simple: it states that measurement outcomes depend only on the physical state λ of the system, not on the preparation procedure which led to λ as the physical state of the system. PBR apparently take this assumption for granted since they do not mention it explicitly.

The assumption of preparation independence is slightly more compli-cated. To formulate it, let us assume that the probability distributions $\mu_{\psi_1}(\lambda)$ and $\mu_{\psi_2}(\lambda)$ have a non-vanishing overlap for distinct quantum states ψ_1 and ψ_2. In this case, as PBR argue, there should be a nonzero minimal probability q for the system to be in λ after either ψ_1 or ψ_2 has been prepared. Now consider a collection of N systems of the same type, all prepared either according to the preparation method associated with the quantum state ψ_1 or according to that associated with the quan-tum state ψ_2. *Preparation independence* is the assumption that there is a probability of at least q^n that the composite (N-part) system is in a phys-ical state λ_c which can come about by *any* of the 2^N possible choices of preparing each system either in ψ_1 or ψ_2.

Given ψ *independence* and *preparation independence*, PBR show how to construct a measurement on the composite system such that, if the probability distributions $\mu_{\psi_1}(\lambda)$ and $\mu_{\psi_2}(\lambda)$ have nontrivial overlap, an outcome will appear with finite probability to which quantum theory ascribes probability 0. Thus, if quantum theory is correct, and if ψ *inde-pendence* and *preparation independence* are true, then the quantum state cannot be 'epistemic' but must be part of the 'ontic' physical state λ, which characterises the system completely.

Schlosshauer and Fine on PBR

Schlosshauer and Fine demonstrate that the assumption of prepara-tion independence in the proof of the PBR theorem may be replaced by the weaker one of *compactness* (see (Schlosshauer and Fine [2012], p. 260404-3)), which still leads to the PBR result. To formulate this assumption, consider two pure quantum states ψ_1 or ψ_2 and an N-component quantum system all components of which are prepared in either ψ_1 or ψ_2, such that the N-component state has the form $|\Psi\rangle = |\psi_{x_1}\rangle \otimes \ldots \otimes |\psi_{x_N}\rangle$ with all the tensor product factors $|\psi_{x_i}\rangle$ being either $|\psi_1\rangle$ or $|\psi_2\rangle$. *Compactness* now states that, if the probability dis-tributions $\mu_{\psi_2}(\lambda)$ and $\mu_{\psi_1}(\lambda)$ overlap, there exists some physical state λ_c of the N-component system which is compatible with all the states $|\Psi\rangle = |\psi_{x_1}\rangle \otimes \ldots \otimes |\psi_{x_N}\rangle$.

Compactness may look innocent and plausible, but, as Schlosshauer and Fine point out, it is in fact 'a strong condition of unifor-mity (like moving from "Everyone has a mother" to "There is a mother of us all")' (Schlosshauer and Fine [2012], pp. 260404–4). Its intuitive appeal, as they argue, translates into intuitive appeal of a 'tracking' composition principle. (Schlosshauer and Fine [2014],

p. 070407-2) But the 'tracking' principle turns out to be violated in many of the best-known existing hidden-variable models, many of which are 'ψ-ontic'. These models are known to be mathematically consistent, and they are not widely considered unnatural in any respects. Thus, the assumption of *compactness* – and a fortiori the stronger one of *preparation independence* – is more assuming than it initially seems, and we should not be surprised if we have to give it up if we want to give a coherent and appealing account of physical states λ that reproduces the quantum predictions. According to Schlosshauer and Fine, the main lesson of the PBR theorem concerns quantum non-separability, not that quantum states are 'real' and hence non-epistemic.

Notes

1. For example, in recent introductory expositions on the interpretation of quantum theory such as (Butterfield [2002]) and (Wallace [2008]) the Copenhagen interpretation is hardly ever mentioned and receives no discussion. A convincing case that the historical dominance of the Copenhagen interpretation was largely contingent and due to the rhetorical gifts of such figures as Bohr and Heisenberg is made by (Cushing [1994]) and (Beller [1999]).
2. Exegetical evidence on whether and, if so, in which guise, the historical proponents of the Copenhagen interpretation accepted collapse is somewhat ambiguous; for more details see Chapter 7.
3. See, for instance, (Maudlin [2011], p. 224) for such considerations, though Maudlin does not ultimately endorse this option.
4. Throughout this work I use 'descriptive' in the vein of 'representing features of reality'. I acknowledge that there are broader uses of 'descriptive' which are no less legitimate.
5. See (Wittgenstein [1978]), Part III in particular.
6. There is a large number of very good introductions to quantum theory. For example, (Landau and Lifshitz [1977]) is an excellent and influential one with a long history.
7. See (Hall [2013], chapter 14) for a rigorous introductory account of the theorem, including a proof, and (Ruetsche [2011], pp. 37–39) for a gentle introduction aimed at philosophers. Precisely speaking, the theorem holds for Hilbert space representations of an exponentiated version of the canonical commutation relations Eq. (2.2), the so-called *Weyl relations*, which are given by

$$\exp(it\mathbf{x})\exp(is\mathbf{p}) = \exp(-i\hbar st)\exp(is\mathbf{p})\exp(it\mathbf{x}), \tag{0.1}$$

where $\mathbf{x} = (x_1, \ldots, x_N)$ and $\mathbf{p} = (p_1, \ldots, p_N)$, for all $s, t \in \mathbb{R}$. The Weyl relations are a formal consequence of the canonical commutation relations, but they avoid problems of the latter that arise from the unboundedness of the operators x_i and p_i.
8. See (Ruetsche [2011]) for a book-length treatment, aimed at philosophers, which covers the issues that arise in quantum theories with infinitely many degrees of freedom in meticulous detail.
9. This terminology is due to (Fine [1973], p. 20).
10. An especially clear version is given, for instance, in a letter from Einstein to Schrödinger, dated 19 June 1935. Einstein complains in that letter that the argument he intended to make had not been made as clear as it should have been in the EPR-paper: 'For reasons of language this was written by Podolsky after several discussions. Still, it did not come out as well as I had originally wanted; rather, the essential thing was, so to speak, smothered by the formalism.' (The letter is dated 19 June 1935, reprinted in (Fine [1996],

p. 35).) For a later version of the argument see Einstein's contribution to (Schilpp [1949]).

11. For a further detailed, self-consciously 'modern', exposition of the problem and its suggested solutions see (Wallace [2008]).

12. See (Struyve [2010]) for a useful and relatively recent overview of the most important achievements.

13. See (Kochen [1985]; Dieks [1998]; Healey [1989]) for some of the most relevant contributions by these authors. For a useful very brief introduction see (Ruetsche [2002], pp. 213–215); for a more detailed overview see (Lombardi and Dieks [2012], section 4).

14. See, for example, (Earman and Ruetsche [2005]) for a helpful discussion of challenges facing modal interpretations of quantum field theories.

15. Section 7.1 offers some considerations on Bohr's position in view of his remarks.

16. See (Penrose [1989], chapter 8) for an accessible introduction to Penrose's idea on quantum gravity's role in collapse.

17. For a defence of this latter approach see (Ghirardi et al. [1995]).

18. See (Dorato and Esfeld [2010]) for critical evaluations of these approaches and suggestions for further options for interpreting GRW theory.

19. Another approach which may be characterised as denying Maudlin's assumption 1.C is the *relational quantum mechanics* approach championed by Rovelli. This approach does acknowledge determinate measurement outcomes, but conceives of them as relational, applying not to the measurement system alone, but only in conjunction with the measuring apparatus. See (Rovelli [1996]) for Rovelli's pioneering work on this interpretation and (Bitbol [2007]) for a comparatively recent criticism and development.

20. Putnam presents this challenge in especially vivid form in (Putnam [2005], section 7).

21. For more detailed and, to my mind, close-to-devastating criticisms of the Everett interpretation see the papers by Kent, Maudlin, and Price in the highly useful anthology (Saunders et al. [2010]).

22. To use Wallace's phrase, see (Wallace [2007], p. 311).

23. See, in particular, (Wittgenstein [1958] § 109). The entire meta-philosophical discussion in that work, especially §§ 109–133, is relevant here.

24. See, for instance, (Wittgenstein [1978], pp. 40, 162, 425).

25. It is notoriously difficult to determine Bohr's views, but it seems that he did not subscribe to the epistemic conception of states. According to Howard, Bohr neither accepted collapse as a proper element of quantum theory nor regarded quantum states as reflecting the epistemic features of the agents who assign them. As he claims, 'Bohr's complementarity interpretation … does not employ wave packet collapse in its account of measurement and does not accord the subjective observer any privileged role in measurement' (See the abstract of (Howard [2004])). However, implicit consent to the epistemic conception of quantum states is attributed to Bohr by Mermin; see (Mermin [2003], p. 521). Bohr's views are discussed in more detail in Chapter 7.

26. See (Peierls [1991], p. 19). Peierls, when he talks of the system 'we are trying to describe', uses the word 'describe' in a broader sense than I do. According

to how I use 'describe', if one claims about states that they 'describe' (physical features of) quantum systems, this implies that for each quantum system there is at most one quantum state which describes it correctly. According to this more narrow usage of 'describe', Peierls would have had to speak of 'the system we are assigning a state to', not 'the system we are trying to describe'.

27. Having in mind this type of approach to quantum theory, Fuchs and Peres speak of the epistemic conception of quantum states, which they favour, as an 'interpretation without interpretation' (Fuchs and Peres [2000], p. 70). Alluding to the same type of approach, Marchildon writes that '[t]he question of the epistemic view [of states] is much the same as the one whether quantum mechanics needs being interpreted' (Marchildon [2004], p. 1454).

28. They correspond essentially to what (Harrigan and Spekkens [2010]) call 'ψ-epistemic' (as opposed to 'ψ-ontic') ontological models.

29. See, in particular, (Fleming [1988]; Myrvold [2002, 2003]).

30. See (Maudlin [2011], pp. 187–195).

31. See (Fuchs [2002], fn. 9 and section 7), and (Timpson [2008], section 2.3).

32. Thus the title of section II of (Fuchs [2010]).

33. For the technical definition of an exchangeable sequence of states see (Caves et al. [2008]), an erratum note to (Caves et al. [2002]).

34. For an analogous result concerning quantum *process* tomography see (Fuchs et al. [2004]).

35. See (Bub [1977, 2007]). In addition, for a proposal of how to interpret Lüders' Rule as the quantum analogue of Bayes' Rule in the context of the quantum Bayesian framework see (Fuchs [2002], section 6). More complicated versions of the measurement update rule are required when dealing with non-sharp measurements or generalised measurements using the formalism of POVMs.

36. Although, as mentioned before, Lüders' Rule can in some special cases replace entropy maximisation, it should be noted that in general Bayesian conditionalisation (i.e., in the quantum context, Lüders' Rule) and entropy maximisation serve different purposes and should not be seen as competing principles. For an instructive assessment of their differing roles see (Jaynes [1988]).

37. See (Shenker [1999]; Henderson [2003]). For an early but still very useful discussion of whether the von Neumann entropy is the adequate quantity in the quantum mechanical context see (Jaynes [1957b]).

38. I would like to thank Richard Healey for a criticism of my earlier account (Friederich [2011]) of the rules of state assignment as constitutive which the current paragraph in the main text is meant to address.

39. See, for instance, (Heisenberg [1958], p. 30).

40. See (Healey [2012a], pp. 742–749) for a discussion of recent experiments on environment-induced decoherence involving fullerene molecules that greatly elaborates and emphasises this point. (Schlosshauer [2005]) gives a helpful introduction to decoherence and clarifies its relevance for the presently most-discussed interpretations of quantum theory.

41. It seems natural to assume that for degenerate observables an (approximately) block-diagonal form of the density matrix assigned should suffice

for licensing application of the Born Rule. More detailed empirical investigations as to what is conceived of as a legitimate application of the Born Rule in quantum theoretical practice would be useful to say more on this matter.

42. The notion was made popular by (Sellars [1953]) and (Brandom [1994]). Section 3 of (Healey [2012a]) gives the essential details of Healey's account of what it means for a NQMC to be licensed by quantum theory.

43. The two complementary conceptions of objectivity alluded to in this passage may be termed the 'invariance' and the 'explicitness' conceptions of objectivity. See (Nozick [2001]) for a book-length study of objectivity as invariance under change of perspective and (Mühlhölzer [1988]) for a detailed take on objectivity as explicitness of all parameters relevant to a statement's truth.

44. See the title of (Ballentine [1970]). Ballentine uses capital letters 'S' and 'I' to indicate that he is referring to a specific interpretation of quantum theory rather than the rather uncontroversial fact that quantum theory makes probabilistic predictions and is in that sense trivially 'statistical'. He attributes the same view to Einstein, Popper, and Blokhintsev (Ballentine [1970], p. 360).

45. Sometimes the notion of propensity is regarded as synonymous with that of objective probability. Here I assume a more narrow notion of propensity according to which propensities are objective probabilities of a specific type.

46. See (Suárez [2007]) for a defence and overview of accounts of quantum probabilities as propensities that contains many useful references.

47. See, for instance, (Gillies [2000]), in particular the classification of probabilities from subjective to objective suggested on pp. 179–180.

48. See (Mellor [1971]) for an earlier work that already contains some of Lewis' central ideas.

49. The analogy to Moore's original sentence '*p*, but I don't believe that *p*' can be made more transparent if the part of the QBMP which reports on the agent's epistemic conditions is put second, just as in Moore's original sentence. (I would like to thank Jeremy Butterfield for pointing this out to me.) A formulation that fulfils this requirement is: 'It is uncertain whether *p*, but I am not uncertain whether *p* (that is, I am absolutely certain that *p*)'.

50. See, for instance, (Beller [1999], p. 2) and the bulk of Part Two of that work.

51. See (Scheibe [1973], p. 19). (Howard [1994]) elaborates on this point in the light of a reading of Bohr according to which the distinction between quantum and classical concepts in Bohr's writing should be seen as corresponding directly to that between pure and mixed quantum states.

52. (Scheibe [1973], pp. 29–35) offers a heroic attempt at coming to terms with this concept.

53. See (Marchildon [2004], p. 1454) for this claim.

54. See (Friederich [2013a], section 6).

55. The theorems due to Bell, Kochen, and Specker are outlined in Appendices A and B. The implications of Gleason's theorem (Gleason [1957]) are similar to those of the Kochen–Specker theorem; see (Held [2013], section 2) for a brief introduction and comparison.

56. The passages which follow in section V of (Fuchs [2010]) offer a detailed argument based on the Kochen–Specker theorem for the view that measurements bring about the values of observables that are usually thought of as brought to light by them.

57. In fact I doubt that quantum Bayesianism can coherently view quantum theory as prescriptive: it denies that there is such a thing as correctly assigning a quantum state and is therefore in tension with the very idea of a *correct* application of quantum theory. It is hard to see how, on this account, quantum theory can have any prescriptive force whatsoever.

58. Quantum Bayesianism concurs that the Born Rule has the normative role of prescribing which probabilities to assign on the basis of which quantum states (Fuchs [2010], p. 8). It denies, however, that the theory has any normative force concerning which states are to be assigned on the basis of what evidence. This denial creates the problems discussed over and over again in the present work.

59. This is the main motivation for Railton's deductive-nomological account of probabilistic explanation (Railton [1978]), which is based, however, on a propensity interpretation of probabilities that does not seem straightforwardly combinable with the account of quantum probabilities given by the Rule Perspective. An alternative approach to the explanation of low probability events departs from Salmon's *statistical relevance* account of probabilistic explanation; see (Salmon [1971]). Salmon's account faces its own problems, which seem avoidable only by accepting aspects of a causal account of explanation. How this may be done in the context of the Rule Perspective is outlined further below in the continuation of the main text.

60. See (Healey [forthcoming], version deposited as http://philsci-archive.pitt. edu/8752/, p. 2).

61. Nuclear fission is the most important technological application of stimulated decay.

62. A further desideratum for quantum theoretical explanations to be satisfying is perhaps that they help us attain a more *unified* understanding of what is to be explained. Spelling out what this means, however, seems no more difficult on an epistemic than an ontic account of quantum states, which is why I neglect this issue for the purposes of the present investigation.

63. See (Fuchs [2010], p. 21, fn. 31) for a quantum Bayesian endorsement of explanatory anti-reductionism and (Timpson [2008], pp. 592, 600) for helpful remarks on the role and importance of anti-reductionism in quantum Bayesianism.

64. For an introduction to Bell's theorem aimed at philosophers see (Shimony [2012]). For a brief sketch, see Appendix A.

65. See section 7 of (Shimony [2012]). An influential argument for peaceful coexistence due to (Jarrett [1984]) has been thoroughly criticised in the past few years, for example by (Maudlin [2011], pp. 85–90). Others have argued that it rests on serious misunderstandings of the points Bell really wanted to make; see (Norsen [2009, 2011]; Näger [2013]).

66. (Maudlin [2011], pp. 141–144) elaborates on this point in great detail. While Maudlin thinks that the quantum correlations require superluminal causation, he does not regard this as by itself raising any serious problems of incompatibility between quantum theory and special relativity. What he

does regard as problematic, however, is quantum theory's violation of the probabilistic criterion given below.

67. Huw Price enthusiastically recommends backward causation as the clue towards resolving the foundational problems of quantum theory; see chapters 8 and 9 of (Price [1996]).

68. See, for instance, (Butterfield [1992]) and (Maudlin [2011], chapter 5). For recent critical voices see (Healey [2012b]) and (Fenton-Glynn and Kroedel [forthcoming]).

69. For a seminal book-length treatment of causation and causal explanation from an interventionist point of view see (Woodward [2003]).

70. See (Suárez [2013]) for a useful recent discussion of whether quantum theory sanctions superluminal causation according to interventionist accounts of causation that addresses this difficulty in detail. Suárez argues that the question remains elusive and reaches no unambiguous conclusion. However, he does not take into account the argument against superluminal causation in quantum theory given by (Healey [2012b]), outlined further below in the main text.

71. See (Woodward [2003], Section 3.5) for support for this claim and discussion of the consequences.

72. See (Healey [2012b], p. 23) for a more detailed version of the argument sketched in what follows.

73. Space-time region 3 in the figure that Bell refers to does not cover the entire backward light cone of region 1, but only a part of it that completely shields region 1 from the backward light cone of region 2 (See region 3 in Figure 2 in Appendix A) The entire backward light cone of region 1 is the most extended such region and contains all the information one may wish to have about region 1's causal past in a locally causal theory. For the sake of simplicity, I will therefore let the backward light cone of region 1 play the role of region 3 in Bell's reasoning.

74. The criterion (BPLC), as formulated here, neglects variables for the measurement settings in regions 1 and 2 since the main point that I wish to make in this chapter can be made without putting much emphasis on them. A more complete formulation of Bell's criterion that includes the measurement settings is given in Eqs. (A.1) and (A.2).

75. For a useful and detailed investigation of criteria related to (BPLC), which focuses on very different aspects than the present discussion, see (Butterfield [2007]).

76. On the side of those who subscribe to a non-ontic account of quantum states, (Healey [forthcoming]) denies the very applicability of (BPLC) to quantum theory. While I sympathise with Healey's view of quantum states, it seems to me that the case for the violation of (BPLC) in quantum theory can be saved by inserting a description of the (objective) joint preparation procedure for the variable E instead of ψ_{EPRB}. If this is so, quantum theory violates (BPLC) independently of whether one grants 'beable status' to the wave function or not.

77. See (Lewis [1986a], pp. 87–89) for details.

78. In what follows I will use the letter 'A' to denote both the proposition concerning some chancy fact as well as the chancy event itself. So, the

expression '*Pr(A)*' can either be read as 'the chance of (the proposition) *A* being true' or as 'the chance of (the event) *A* occurring'.

79. According to Ned Hall (Hall [1994, 2004]), it is possible to formulate the Principal Principle without recourse to the notion of admissible evidence by giving an explicit formal definition of admissibility. However, as argued in (Friederich [forthcoming]), this definition does not help one settle the question of what evidence is to be counted as admissible in the quantum context, which means that the Hallian approach to the Principal Principle differs merely in terminology, not in consequences, from the Lewisian one adopted here. See (Friederich [forthcoming], section 3) for more details.

80. See (Lewis [1986a], p. 97) as 'the Principal Principle reformulated'.

81. See (Gell-Mann et al. [1954]) for a historically important reference. See (Haag [1993], pp. 57, 107) for a version in the language of the mathematically rigorous algebraic approach to quantum theories.

82. See (Eberhard [1978]; Ghirardi et al. [1980]). Bell's own criticism of no-(superluminal) signalling as the appropriate expression of relativistic causal space-time structure is addressed in Section 10.6.

83. See the abstract of (Gisin [2009]). The intuition which underlies the criticism developed here underlies all the famous Gedanken experiments which are meant to highlight what is mysterious about EPR correlations. They are formulated, for instance, in terms of agents who are asked various questions in isolation from each other and have to answer them such that correlations analogous to the EPR correlations come about in their answers. It is easily shown that there are no strategies by means of which the agents could achieve this without being able to communicate after being asked the questions. In the argument presented here, *nature herself*, or the systems being measured, play the role of the agents in these Gedanken experiments, who are trying to produce the quantum correlations in their answers.

84. See (Price [1996], p. 12). For more details and a meticulous defence of the block universe view see (the whole of) (Price [1996]), in particular chapter 4. For a recent criticism, in part directed explicitly against Price, see (Maudlin [2007], chapter 4). Some philosophers, for instance (Dieks [2006]), argue that the conception of objective *local* (rather than *global*) becoming (and, so, of an objective *local* flow of time) is perfectly compatible with what they regard as the essential constituents of the block universe view. Views which acknowledge objective local becoming are not versions of the block universe view in the sense assumed here.

    As far as I can see, the strongest argument for the non-objectivity of the flow of time (or objective becoming, whether global or local) rests on the fact that there might be regions in our universe (or multiverse) where the entropy gradient has the opposite sign to where we live. This strongly suggests that either for us or for hypothetical (not unlikely real) creatures in those other regions the subjective sense of time is opposed to the direction of objective becoming, an idea which seems artificial and unappealing and speaks against the idea of an objective flow of time itself. See (Maudlin [2007], pp. 271–274) for a criticism of this argument and (Price [2011], section 3.10), for a, to my mind convincing, response.

85. An analogous double standard in the orthodox view of the EPR-scenario has recently been pointed out along different lines in (Evans et al. [2013]).

86. See, for instance, (Strocchi [2013]) for an extremely useful overview and collection of results on quantum field theories with empirical applications from a rigorous viewpoint, using many concepts and techniques of the algebraic approach.

87. The exchange of arguments between (Fraser [2011]) and (Wallace [2011]) may serve as an overview of the most important aspects in this debate.

88. See (Teller [1995], chapter 7) for an introduction to the conceptual challenges raised by renormalisation that is aimed at philosophers.

89. An algebra is a vector space with a bilinear vector product defined on it. A C^* algebra is an algebra that is isomorphic to a sub-algebra of some algebra $\mathcal{B}(\mathcal{H})$ of bounded linear operators on a Hilbert space $\mathcal{H}$ that is closed with respect to the adjoint operation $*$.

90. A sequence A_n of operators on $\mathcal{H}$ is said to converge in the 'strong topology' iff $|(A_n - A)|\psi\rangle| \mapsto 0$ as $n \mapsto \infty$ for each $|\psi\rangle \in \mathcal{H}$. The 'strong closure' of $\pi(\mathcal{A})$ includes all elements of $\pi(\mathcal{A})$ itself, together with the limits of all sequences of $\pi(\mathcal{A})$ that are convergent in this topology.

91. See, for instance, (Callender [2001]) for considerations on which this objection might be based.

92. Ruetsche defends this view by drawing attention to the success of renormalisation group theory in accounting for universality and critical phenomena, which success, as she argues, is crucially based on the idealisations of the thermodynamic limit (see (Ruetsche [2011], p. 339)). This claim can be questioned, however, for instance by pointing to the fact that analyses employing renormalisation group techniques may still have to take into account the finiteness of real systems to enjoy maximal explanatory success.

93. Chapter 13 of (Ruetsche [2011]) provides a useful introduction to spontaneous symmetry breaking that is aimed at philosophers, and gives a rigorous definition; see p. 300.

94. See, for instance, (Halzen and Martin [1984], chapter 14) for a standard account of the Higgs mechanism in terms of this notion. A *gauge symmetry*, at least to a first and approximate characterisation of the notion, is one where different variable configurations related by symmetry transformations correspond to one and the same physical situation.

95. The main advantage of the unitary gauge is that it is useful in a classical argument giving an idea as to which degrees of freedom are *physical* and which can be 'gauged away'.

96. See (Friederich [2013b]) for an interpretive discussion of this issue that is aimed at philosophers.

97. See, for instance, (Linde [1979]) for a classic reference that gives a first broad impression of this phase structure.

98. See Eqs. (A.1) and (A.2) for the full criterion that includes the variables for the apparatus settings.

99. See (Schlosshauer and Fine [2012, 2014]) for discussion and criticism of the assumptions used in the PBR theorem, *preparation independence* in particular. A short summary of some of the criticisms by Schlosshauer and Fine can be found at the end of Appendix C.

100. See (Laudisa [2014]) for a recent criticism of what he terms the 'no-go philosophy of quantum mechanics', arguing that the significance of the no-go

theorems is in general rather limited due to the controversial character of the assumptions on which each of them, respectively, rests.

101. See (Shimony [2012]) for a more detailed outline of the theorem and its ramifications, and the references therein for an overview of the history of the theorem and its developments.

102. See (Seevinck and Uffink [2011]) for the details of what exactly should be meant here by 'complete' in the technical sense and how exactly the equations (A.1) and (A.2) are to be interpreted for the sake of a formulation that is 'sufficiently sharp and clean for mathematics' (Bell [2004], p. 239).

103. See (Redhead [1987], p. 121) for this terminology

104. See (Redhead [1987], p. 18) for this terminology.

Bibliography

D. Z Albert and R. Galchen. Was Einstein wrong?: A quantum threat to special relativity. *Scientific American*, 300:32–39, 2009.

A. Arageorgis. *Fields, Particles and Curvature: Foundations and Philosophical Aspects of Quantum Field Theory on Curved Spacetime and the Algebraic Approach to QFT*. PhD thesis, University of Pittsburgh, 1995.

A. Aspect, P. Grangier, and G. Roger. Experimental realization of Einstein-Podolsky-Rosen-Bohm *Gedankenexperiment*: A new violation of Bell's inequalities. *Physical Review Letters*, 49:91–94, 1982.

L. Ballentine. The Statistical Interpretation of quantum mechanics. *Reviews of Modern Physics*, 42:358–381, 1970.

J. S. Bell. *Speakable and Unspeakable in Quantum Mechanics*. Cambridge: Cambridge University Press, 2nd edition, 2004.

J. S. Bell. Against 'measurement'. *Physics World*, August:33–40, 2007.

M. Beller. *Quantum Dialogue: The Making of a Revolution*. Chicago: University of Chicago Press, 1999.

M. Bitbol. Physical relations or functional relations? A non-metaphysical construal of Rovelli's relational quantum mechanics. 2007. http://philsci-archive.pitt.edu/3506/.

D. Bohm and B. Hiley. *The Undivided Universe: An Ontological Interpretation of Quantum Theory*. London: Routledge, 1993.

N. Bohr. *Atomic Theory and the Description of Nature*. Cambridge: Cambridge University Press, 1934.

N. Bohr. Quantum physics and philosophy—causality and complementarity. In *Essays 1958-1962 on Atomic Physics and Human Knowledge*, pages 1–7. New York: Interscience, 1962.

R. Brandom. *Making It Explicit*. Cambridge, MA: Harvard University Press, 1994.

J. Bub. Von Neumann's projection postulate as a probability conditionalization rule in quantum mechanics. *Journal of Philosophical Logic*, 6: 381–390, 1977.

J. Bub. Quantum probabilities as degrees of belief. *Studies in History and Philosophy of Modern Physics*, 38:232–254, 2007.

J. Butterfield. David Lewis meets John Bell. *Philosophy of Science*, 59:26–43, 1992.

J. Butterfield. Some worlds of quantum theory. In R. J. Russell, N. Murphy, and C. J. Isham, editors, *Quantum Physics and Divine*

Actions. Vatican Observatory Publications, 2002. http://philsci-archive.pitt.edu/203/.

J. Butterfield. Stochastic Einstein locality revisited. *British Journal for the Philosophy of Science*, 58:805–867, 2007.

A. Cabello, J. Estebaranz, and G. Garcìa-Alcaine. Bell-Kochen-Specker theorem: A proof with 18 vectors. *Physics Letters A*, 212:183–187, 1996.

C. Callender. Taking thermodynamics too seriously. *Studies in History and Philosophy of Modern Physics*, 32:539–554, 2001.

K. Camilleri. *Heisenberg and the Interpretation of Quantum Mechanics: The Physicist as a Philosopher*. Cambridge: Cambridge University Press, 2009.

C. M. Caves, C. A. Fuchs, and R. Schack. Unknown quantum states: The quantum de Finetti representation. *Journal of Mathematical Physics*, 43: 4537–4559, 2002.

C. M. Caves, C. A. Fuchs, and R. Schack. Erratum: Unknown quantum states: The quantum de Finetti representation. *Journal of Mathematical Physics*, 49:019902, 2008.

J. Cohen and C. Callender. A better best system account of lawhood. *Philosophical Studies*, 145:1–34, 2009.

J. Conway and S. Kochen. The free will theorem. *Foundations of Physics*, 36:1441–1473, 2006.

J. T. Cushing. *Quantum Mechanics: Historical Contingency and the Copenhagen Hegemony*. Chicago: University of Chicago Press, 1994.

B. d'Espagnat. *Conceptual Foundations of Quantum Mechanics*. Reading, Mass.: Addison-Wesley, 2nd edition, 1976.

D. Deutsch. Comment on Lockwood. *British Journal for the Philosophy of Science*, 47:222–228, 1996.

D. Deutsch. Quantum theory of probability and decisions. *Proceedings of the Royal Society of London A*, 455:3129–3137, 1999.

D. Dieks. Becoming, relativity and locality. In D. Dieks, editor, *The Ontology of Spacetime, Vol. 1*, pages 157–176. Amsterdam: Elsevier. 2006.

D. Dieks. The formalism of quantum theory: an objective description of reality? *Annalen der Physik*, 500:174–190, 1988.

M. Dorato and M. Esfeld. GRW as an ontology of dispositions. *Studies in History and Philosophy of Modern Physics*, 41:41–49, 2010.

J. Earman and L. Ruetsche. Relativistic invariance and modal interpretations. *Philosophy of Science*, 72:557–583, 2005.

P. H. Eberhard. Bell's theorem and the different concepts of locality. *Nuovo Cimento*, 46B:392–419, 1978.

A. Einstein, B. Podolski, and N. Rosen. Can quantum mechanical description of physical reality be considered complete? *Physical Review*, 47:777–780, 1935.

S. Elitzur. Impossibility of spontaneously breaking local symmetries. *Physical Review D*, 12:3978–3982, 1975.

P. W. Evans, H. Price, and K. B. Wharton. New slant on the EPR-Bell experiment. *British Journal for the Philosophy of Science*, 64:297–324, 2013.

H. Everett. 'Relative state' formulation of quantum mechanics. *Reviews of Modern Physics*, 29:454–462, 1935. reprinted in ?, pp. 315-323.

J. Faye. Copenhagen interpretation of quantum mechanics. In E. N. Zalta, editor, *The Stanford Encyclopedia of Philosophy*. Fall 2008 edition, 2008. http://plato.stanford.edu/archives/fall2008/entries/qm-copenhagen/.

L. Fenton-Glynn and T. Kroedel. Relativity, quantum entanglement, counterfactuals, and causation. *British Journal for the Philosophy of Science*, forthcoming. doi: 10.1093/bjps/axt040.

A. Fine. Probability and the interpretation of quantum mechanics. *British Journal for the Philosophy of Science*, 24:1–37, 1973.

A. Fine. *The Shaky Game: Einstein, Realism and the Quantum Theory*. Chicago: University of Chicago Press, 2nd edition, 1996.

G. N. Fleming. Lorentz invariant state reduction, and localization. In A. Fine and M. Forbes, editors, *PSA 1988, Vol. 2*, pages 112–126. East Lansing, MI: Philosophy of Science Association, 1988.

D Fraser. How to take particle physics seriously: A further defence of axiomatic quantum field theory. *Studies in History and Philosophy of Modern Physics*, 42:126–135, 2011.

S. Friederich. How to spell out the epistemic conception of quantum states. *Studies in History and Philosophy of Modern Physics*, 42:149–157, 2011.

S. Friederich. Interpreting Heisenberg interpreting quantum states. *Philosophia Naturalis*, 50:85–114, 2013a.

S. Friederich. Gauge symmetry breaking in gauge theories—in search of clarification. *European Journal for Philosophy of Science*, 3:157–182, 2013b.

S. Friederich. Re-thinking local causality. forthcoming.

C. A. Fuchs. Quantum mechanics as quantum information (and only a little more). In A. Khrennikov, editor, *Quantum Theory: Reconsideration of Foundations*, pages 463–543. Växjö: Växjö University Press, 2002. arXiv:quant-ph/0205039.

C. A. Fuchs. QBism, the perimeter of quantum Bayesianism. 2010. arXiv:1003.5209.

C. A. Fuchs and A. Peres. Quantum theory needs no 'interpretation'. *Physics Today*, 53:70–71, 2000.

C. A. Fuchs and R. Schack. Quantum-Bayesian coherence. 2009. arXiv:0906.2187.

C. A. Fuchs, R. Schack, and P. F. Scudo. De Finetti representation theorem for quantum-process tomography. *Physical Review A*, 69:062305, 2004.

M. Gell-Mann, M. L. Goldberger, and W. E. Thirring. Use of causality conditions in quantum theory. *Physical Review*, 95:1612–1627, 1954.

G. Ghirardi. Collapse theories. In E. N. Zalta, editor, *The Stanford Encyclopedia of Philosophy*. Winter 2011 edition, 2011. http://plato.stanford.edu/archives/win2011/entries/qm-collapse/.

G. C. Ghirardi, A. Rimini, and T. Weber. A general argument against superluminal transmission through the quantum mechanical measurement process. *Lettere al Nuovo Cimento*, 27:293–298, 1980.

G. C. Ghirardi, R. Grassi, and F. Benatti. Describing the macroscopic world: Closing the circle within the dynamical reduction program. *Foundations of Physics*, 25:5–38, 1995.

D. Gillies. *Philosophical Theories of Probability*. London: Routledge, 2000.

N. Gisin. Quantum nonlocality: How does nature do it? *Science*, 326: 1357–1358, 2009.

A.M. Gleason. Measures on the closed subspaces of a Hilbert Space. *Journal of Mathematics and Mechanics*, 6:885–893, 1957.

R. Haag. *Local Quantum Physics*. Berlin: Springer, 1993. corrected edition.

A. Hájek. Fifteen arguments against hypothetical frequentism. *Erkenntnis*, 70:211–235, 2009.

A. Hájek and N. Hall. Induction and probability. In P. Machamer and M. Silberstein, editors, *The Blackwell Guide to the Philosophy of Science*, pages 149–172. Oxford: Blackwell, 2002.

B C. Hall. *Quantum Theory for Mathematicians*. New York, Heidelberg, Dordrecht, London: Springer, 2013.

N. Hall. Correcting the guide to objective chance. *Mind*, 103:505–517, 1994.

N. Hall. Two mistakes about credence and chance. *Australasian Journal of Philosophy*, 82:93–111, 2004.

F. Halzen and A. D. Martin. *Quarks and Leptons: An Introductory Course in Modern Particle Physics*. New York: John Wiley & Sons, 1984. corrected edition.

L. Hardy. Quantum ontological excess baggage. *Studies in History and Philosophy of Modern Physics*, 35:267–276, 2004.

N. Harrigan and R. W. Spekkens. Einstein, incompleteness, and the epistemic view of quantum states. *Foundations of Physics*, 40:125–157, 2010.

R. Healey. *The Philosophy of Quantum Mechanics: An Interactive Interpretation*. Cambridge: Cambridge University Press, 1989.

R. Healey. Quantum theory: a pragmatist approach. *British Journal for the Philosophy of Science*, 63:729–771, 2012a.

R. Healey. How to use quantum theory locally to explain 'non-local' correlations. 2012b. http://philsci-archive.pitt.edu/8864/.

R. Healey. How quantum theory helps us explain. *British Journal for the Philosophy of Science*, forthcoming. doi: 10.1093/bjps/axt031.

W. Heisenberg. Über den anschaulichen Inhalt der quantentheoretischen Kinematik und Mechanik. *Zeitschrift für Physik*, 43:172–198, 1927.

W. Heisenberg. *Physics and Philosophy: The Revolution in Modern Science*. London: George Allen & Unwin, 1958. repr. 2007 by Harper, New York, page numbers referring to this edition.

W. Heisenberg. *Philosophical Problems of Quantum Physics*. Woodbridge Conn: Ox Bow, 1979. Reprint of *Philosophical Problems of Nuclear Science*, New York: Pantheon, 1952.

C. Held. The Kochen-Specker Theorem. In E. N. Zalta, editor, *The Stanford Encyclopedia of Philosophy*. Spring 2013 edition, 2013. http://plato.stanford.edu/archives/spr2013/entries/kochen-specker/.

L. Henderson. The von Neumann entropy: A reply to Shenker. *British Journal for the Philosophy of Science*, 54:291–296, 2003.

C. Hoefer. The third way on objective probability: a sceptic's guide to objective chance. *Mind*, 116:549–596, 2007.

P. Holland. *The Quantum Theory of Motion: An Account of the de Broglie-Bohm Causal Interpretation of Quantum Mechanics*. Cambridge: Cambridge University Press, 1993.

P. Horwich. Wittgensteinian Bayesianism. *Midwest Studies in Philosophy*, 18:62–75, 1993. repr. in Curd, M. and Cover, J. A. (eds.), *Philosophy of Science: The Central Issues*, London: W. W. Norton & Company Ltd., 1998, pp. 607-624, page numbers referring to this edition.

D. Howard. What makes a classical concept classical? Toward a reconstruction of Niels Bohr's philosophy of physics. In *Niels Bohr and Contemporary Philosophy, vol. 158 of Boston Studies in the Philosophy of Science*, pages 201–229. Dordrecht: Kluwer, 1994.

D. Howard. Who invented the "Copenhagen Interpretation"? A study in mythology. *Philosophy of Science*, 71:669–682, 2004.

J. Jarrett. On the physical significance of the locality conditions in the Bell arguments. *Nous*, 18:569–589, 1984.

E. T. Jaynes. Information theory and statistical mechanics. *The Physical Review*, 106:620–630, 1957a.

E. T. Jaynes. Information theory and statistical mechanics II. *The Physical Review*, 108:171.190, 1957b.

E. T. Jaynes. The relation of Bayesian and maximum entropy methods. *IEEE Transactions on Systems Science and Cybernetics*, 4:227–241, 1988.

P. Jordan. Quantenphysikalische Bemerkungen zur Biologie und Psychologie. *Erkenntnis*, 4:215–252, 1934.

S. Kochen. A new interpretation of quantum mechanics. In P. Mittelstaedt and P. Lahti, editors, *Symposium on the Foundations of Modern Physics 1985*, pages 151–169. Singapore: World Scientific, 1985.

S. Kochen and E. Specker. The problem of hidden variables in quantum mechanics. *Journal of Mathematics and Mechanics*, 17:59–87, 1967.

L. D. Landau and E. M. Lifshitz. *Quantum Mechanics: Non-relativistic Theory*. Oxford: Pergamon Press, 3rd edition, 1977. English translation by J. B. Sykes and J. S. Bell.

F. Laudisa. Against the no-go philosophy of quantum mechanics. *European Journal for Philosophy of Science*, 4:1–17, 2014.

A. Leggett. Nonlocal hidden-variable theories and quantum mechanics: an incompatibility theorem. *Foundations of Physics*, 33:1469–1493, 2003.

D. K. Lewis. Causation. *Journal of Philosophy*, 70:556–567, 1973. reprinted in: "Philosophical Papers, Vol. II", New York: Oxford University Press, 1986.

D. K. Lewis. A subjectivists's guide to objective chance. In *Philosophical Papers, Vol. II*, pages 83–132. New York: Oxford University Press, 1986a. originally published in: *Studies in Inductive Logic and Probability, Vol. II*, ed. by Jeffrey, R. C., Berkeley: University of California Press, 1980.

D. K. Lewis. Causal explanation. In *Philosophical Papers, Vol. II*, pages 214–240. New York: Oxford University Press, 1986b.

D. K. Lewis. Humean supervenience debugged. *Mind*, 103:473–490, 1994.

A. D. Linde. Phase transitions in gauge theories and cosmology. *Reports on Progress in Physics*, 42:389–437, 1979.

O. Lombardi and D. Dieks. Modal interpretations of quantum mechanics. In E. N. Zalta, editor, *The Stanford Encyclopedia of Philosophy*. Winter 2012 edition, 2012. http://plato.stanford.edu/archives/win2012/entries/qm-modal/.

F. London and E. Bauer. *La théorie de l'observation en mécanique quantique*. Paris: Hermann, 1939. English translation: "The theory of observation in quantum mechanics" in ?, pp. 217-259, page numbers referring to this edition.

L. Marchildon. Why should we interpret quantum mechanics? *Foundations of Physics*, 34:1453–1466, 2004.

T. Maudlin. Three measurement problems. *Topoi*, 14:7–15, 1995.

T. Maudlin. *The Metaphysics Within Physics*. Oxford: Oxford University Press, 2007.

T. Maudlin. *Quantum Theory and Relativity Theory: Metaphysical Intimations of Modern Physics*. Oxford: Wiley-Blackwell, 3rd edition, 2011.

H. Mellor. *The Matter of Chance*. Cambridge: Cambridge University Press, 1971.

P. C. Menzies and H. Price. Causation as a secondary quality. *British Journal for the Philosophy of Science*, 44:187–203, 1993.

N. D. Mermin. Copenhagen computation. *Studies in History and Philosophy of Modern Physics*, 34:511–522, 2003.

F. Mühlhölzer. On objectivity. *Erkenntnis*, 28:185–230, 1988.

W. C. Myrvold. On peaceful coexistence: is the collapse postulate incompatible with relativity? *Studies in History and Philosophy of Modern Physics*, 33:435–466, 2002.

W. C. Myrvold. Relativistic quantum becoming. *British Journal for the Philosophy of Science*, 54:475–500, 2003.

P. Näger. A stronger Bell argument for quantum non-locality. 2013. http://philsci-archive.pitt.edu/9068/.

T. Norsen. Local causality and completeness: Bell vs. Jarrett. *Foundations of Physics*, 39:273–294, 2009. http://philsci-archive.pitt.edu/9068/.

T. Norsen. John S. Bell's concept of local causality. *American Journal of Physics*, 79:1261–1275, 2011. http://philsci-archive.pitt.edu/9068/.

R. Nozick. *Invariances: The Structure of the Objective World*. Cambridge MA: Belknap Press, 2001.

R. Peierls. In defence of 'measurement'. *Physics World*, 4:19–21, 1991.

R. Penrose. *The Emperor's New Mind*. Oxford: Oxford University Press, 1989.

A. Peres. Two simple proofs of the Kochen-Specker theorem. *Journal of Physics A*, 24:L175–L178, 1991.

R. Popper, K. The propensity interpretation of probability. *British Journal for the Philosophy of Science*, 10:25–42, 1959.

H. Price. *Time's Arrow and Archimedes' Point: New Directions for the Physics of Time*. Oxford, New York: Oxford University Press, 1996.

H. Price. The flow of time. In C. Callender, editor, *The Oxford Handbook of Philosophy of Time*, pages 276–311. Oxford: Oxford University Press, 2011.

H. Price. Causation, chance, and the rational significance of supernatural evidence. *Philosophical Review*, 121:483–538, 2012.

M. F. Pusey, J. Barrett, and T. Rudolph. On the reality of the quantum state. *Nature Physics*, 8:476–479, 2012.

H. Putnam. A philosopher looks at quantum mechanics (again). *British Journal for the Philosophy of Science*, 56:615–634, 2005.

P. Railton. A deductive-nomological model of probabilistic explanation. *Philosophy of Science*, 45:206–226, 1978.

M. Redhead. *Incompleteness, Nonlocality, and Realism. A Prolegomenon to the Philosophy of Quantum Mechanics*. Oxford: Clarendon Press, 1987.

H. P. Robertson. The uncertainty principle. *Physical Review*, 34:163–164, 1929.

C. Rovelli. Relational quantum mechanics. *International Journal of Theoretical Physics*, 35:1637–1678, 1996.

L. Ruetsche. Interpreting quantum theories. In P. Machamer and M. Silberstein, editors, *The Blackwell Guide to the Philosophy of Science*, pages 199–226. Oxford: Blackwell, 2002.

L. Ruetsche. *Interpreting Quantum Theories*. Oxford: Oxford University Press, 2011.

W. Salmon. Statistical explanation. In W. Salmon, editor, *Statistical Explanation and Statistical Relevance*, pages 29–87. Pittsburgh: University of Pittsburgh Press, 1971.

S. Saunders, J. Barrett, A. Kent, and D. Wallace, editors. *Many Worlds? Everett, Quantum Theory and Reality*, 2010. Oxford: Oxford University Press.

E. Scheibe. *The Logical Analysis of Quantum Mechanics*. Oxford: Pergamon Press, 1973.

P. A. Schilpp, editor. *Albert Einstein: Philosopher-Scientist*, 1949. La Salle, IL: Open Court.

M. Schlosshauer. Decoherence, the measurement problem, and interpretations of quantum mechanics. *Reviews of Modern Physics*, 76: 1267–1305, 2005.

M. Schlosshauer and A. Fine. Implications of the Pusey-Barrett-Rudolph quantum no-go theorem. *Physical Review Letters*, 108:260404, 2012.

M. Schlosshauer and A. Fine. No-go theorem for the composition of quantum systems. *Physical Review Letters*, 112:070407, 2014.

E. Schrödinger. Discussion of probability relations between separated systems. *Mathematical Proceedings of the Cambridge Philosophical Society*, 31:555–563, 1935.

J. Searle. *Speech Acts*. New York: Cambridge University Press, 1969.

M. Seevinck. Can quantum theory and special relativity peacefully coexist? 2010. http://arxiv.org/abs/1010.3714.

M. Seevinck and J. Uffink. Not throwing out the baby with the bathwater: Bell's condition of local causality mathematically 'sharp and clean'. In D. Dieks, W.J. Gonzalez, S. Hartmann, T. Uebel, and M. Weber, editors, *Explanation, Prediction and Confirmation. New Trends and Old Ones Reconsidered*, pages 425–450. Dordrecht: Springer, 2011.

W. Sellars. Inference and meaning. *Mind*, 62:313–338, 1953.

O. Shenker. Is $-k\mathrm{Tr}(\rho \log \rho)$ the entropy in quantum mechanics? *British Journal for the Philosophy of Science*, 50:33–48, 1999.

A. Shimony. Metaphysical problems in the foundations of quantum mechanics. *International Philosophical Quarterly*, 18:3–17, 1978.

A. Shimony. Bell's theorem. In E. N. Zalta, editor, *The Stanford Encyclopedia of Philosophy*. Winter 2012 edition, 2012. http://plato.stanford.edu/archives/win2012/entries/bell-theorem/.

R. W. Spekkens. Evidence for the epistemic view of quantum states: A toy theory. *Physical Review A*, 75:032110, 2007.

H. P. Stapp. *Mind, Matter, and Quantum Mechanics*. Berlin, Heidelberg: Springer, 2nd edition, 2003.

D. Stoljar and N. Damnjanovic. The deflationary theory of truth. In E. N. Zalta, editor, *The Stanford Encyclopedia of Philosophy*. Summer 2012 edition, 2012. http://plato.stanford.edu/archives/sum2012/entries/truth-deflationary/.

F. Strocchi. *Symmetry Breaking*. Berlin, Heidelberg: Springer, 2nd edition, 2008.

F. Strocchi. *An Introduction to Non-Perturbative Foundations of Quantum Field Theory*. Oxford: Oxford University Press, 2013.

W. Struyve. Pilot-wave theory and quantum fields. *Reports on Progress in Physics*, 73:106001, 2010.

M. Suárez. Quantum propensities. *Studies in History and Philosophy of Modern Physics*, 38:418–438, 2007.

M. Suárez. Interventions and causality in quantum mechanics. *Erkenntnis*, 78:199–213, 2013.

W. W. Tait. Beyond the axioms: the question of objectivity in mathematics. *Philosophia Mathematica*, 9:21–36, 2001.

P. Teller. *An Interpretive Introduction to Quantum Field Theory*. Princeton: Princeton University Press, 1995.

C. G. Timpson. Quantum Bayesianism: A study. *Studies in History and Philosophy of Modern Physics*, 39:579–609, 2008.

R. Tumulka. A relativistic version of the Ghirardi-Rimini-Weber model. *Journal of Statistical Physics*, 125:821–840, 2006.

L. Vaidman. On schizophrenic experiences of the neutron or why we should believe in the many-worlds interpretation of quantum theory. *International Studies in the Philosophy of Science*, 12:245–261, 1998.

A. Valentini. On Galilean and Lorentz Invariance in Pilot-Wave Dynamics. *Physics letters A*, 228:215–222, 1997.

B. C. van Fraassen. The Einstein-Podolsky-Rosen paradox. *Synthese*, 29: 291–309, 1974.

J. von Neumann. *Mathematical Foundations of Quantum Mechanics*. Princeton, NJ: Princeton University Press, 1955. First published in German in 1932: "Mathematische Grundlagen der Quantenmechanik", Berlin: Springer.

D. Wallace. Quantum probability from subjective likelihood: Improving on Deutsch's proof of the probability rule. *Studies in History and Philosophy of Modern Physics*, 38:311–332, 2007.

D. Wallace. Philosophy of quantum mechanics. In D. Rickles, editor, *The Ashgate companion to contemporary philosophy of physics*, pages 16–98. Aldershot: Ashgate, 2008.

D. Wallace. Taking particle physics seriously: A critique of the algebraic approach to quantum field theory. *Studies in History and Philosophy of Modern Physics*, 42:116–125, 2011.

E. P. Wigner. *Symmetries and Reflections: Scientific Essays of Eugene P. Wigner*. Cambridge, MA: MIT Press, 1967.

L. Wittgenstein. *Philosophical Investigations*. Oxford: Blackwell, 2nd edition, 1958.

L. Wittgenstein. *Remarks on the Foundations of Mathematics*. Oxford: Blackwell, 3rd edition, 1978.

J. Woodward. The causal mechanical model of explanation. In W. Salmon and P. Kitcher, editors, *Minnesota Studies in the Philosophy of Science, Vol 13: Scientific Explanation*, pages 357–383. Minneapolis: University of Minnesota Press, 1989.

J. Woodward. *Making Things Happen: A Theory of Causal Explanation*. Oxford: Oxford University Press, 2003.

H. Zbinden, J. Brendel, N. Gisin, and W. Tittel. Experimental test of nonlocal quantum correlations in relativistic configurations. *Physical Review A*, 63:022111, 2001.

Index

Printed in the United States
by Baker & Taylor Publisher Services